EXPLORING THE INTERSECTION OF
SCIENCE EDUCATION AND 21ST CENTURY SKILLS

A Workshop Summary

Margaret Hilton, *Rapporteur*

Board on Science Education
Center for Education

Division of Behavioral and Social Sciences and Education

NATIONAL RESEARCH COUNCIL
OF THE NATIONAL ACADEMIES

THE NATIONAL ACADEMIES PRESS
Washington, D.C.
www.nap.edu

THE NATIONAL ACADEMIES PRESS 500 Fifth Street, N.W. Washington, DC 20001

NOTICE: The project that is the subject of this report was approved by the Governing Board of the National Research Council, whose members are drawn from the councils of the National Academy of Sciences, the National Academy of Engineering, and the Institute of Medicine. The members of the committee responsible for the report were chosen for their special competences and with regard for appropriate balance.

This study was supported by Award No. N01-OD-4-2139 TO#199 between the National Academy of Sciences and the National Institutes of Health Office of Science Education and an award between the National Academy of Sciences and the Partnership for 21st Century Skills. The following members of the Partnership for 21st Century Skills contributed to the award: Apple, Inc.; Intel, Inc.; National Education Association; Pearson PLC; and SAP (no longer a member). Additional funding was provided by a contract between the National Academy of Sciences and Ford Motor Company Fund, a member of the Partnership for 21st Century Skills. Any opinions, findings, conclusions, or recommendations expressed in this publication are those of the author(s) and do not necessarily reflect the views of the organizations or agencies that provided support for the project.

International Standard Book Number-13: 978-0-309-14518-3
International Standard Book Number-10: 0-309-14518-X

Additional copies of this report are available from National Academies Press, 500 Fifth Street, N.W., Lockbox 285, Washington, DC 20055; (800) 624-6242 or (202) 334-3313 (in the Washington metropolitan area); Internet, http://www.nap.edu

Copyright 2010 by the National Academy of Sciences. All rights reserved.

Printed in the United States of America

Suggested citation: National Research Council. (2010). *Exploring the Intersection of Science Education and 21st Century Skills: A Workshop Summary.* Margaret Hilton, Rapporteur. Board on Science Education, Center for Education, Division of Behavioral and Social Sciences and Education. Washington, DC: The National Academies Press.

THE NATIONAL ACADEMIES
Advisers to the Nation on Science, Engineering, and Medicine

The **National Academy of Sciences** is a private, nonprofit, self-perpetuating society of distinguished scholars engaged in scientific and engineering research, dedicated to the furtherance of science and technology and to their use for the general welfare. Upon the authority of the charter granted to it by the Congress in 1863, the Academy has a mandate that requires it to advise the federal government on scientific and technical matters. Dr. Ralph J. Cicerone is president of the National Academy of Sciences.

The **National Academy of Engineering** was established in 1964, under the charter of the National Academy of Sciences, as a parallel organization of outstanding engineers. It is autonomous in its administration and in the selection of its members, sharing with the National Academy of Sciences the responsibility for advising the federal government. The National Academy of Engineering also sponsors engineering programs aimed at meeting national needs, encourages education and research, and recognizes the superior achievements of engineers. Dr. Charles M. Vest is president of the National Academy of Engineering.

The **Institute of Medicine** was established in 1970 by the National Academy of Sciences to secure the services of eminent members of appropriate professions in the examination of policy matters pertaining to the health of the public. The Institute acts under the responsibility given to the National Academy of Sciences by its congressional charter to be an adviser to the federal government and, upon its own initiative, to identify issues of medical care, research, and education. Dr. Harvey V. Fineberg is president of the Institute of Medicine.

The **National Research Council** was organized by the National Academy of Sciences in 1916 to associate the broad community of science and technology with the Academy's purposes of furthering knowledge and advising the federal government. Functioning in accordance with general policies determined by the Academy, the Council has become the principal operating agency of both the National Academy of Sciences and the National Academy of Engineering in providing services to the government, the public, and the scientific and engineering communities. The Council is administered jointly by both Academies and the Institute of Medicine. Dr. Ralph J. Cicerone and Dr. Charles M. Vest are chair and vice chair, respectively, of the National Research Council.

www.national-academies.org

PLANNING COMMITTEE ON EXPLORING THE INTERSECTION OF SCIENCE EDUCATION AND THE DEVELOPMENT OF 21ST CENTURY SKILLS

ARTHUR EISENKRAFT (*Chair*), Graduate College of Education, University of Massachusetts, Boston
WILLIAM BONVILLIAN, Washington, DC Office, Massachusetts Institute of Technology
MARCIA C. LINN, Graduate School of Education, University of California, Berkeley
CHRISTINE MASSEY, Institute for Research in Cognitive Science, University of Pennsylvania
CARLO PARRAVANO, Merck Institute for Science Education, Rahway, New Jersey
WILLIAM SANDOVAL, Division of Psychological Studies, University of California, Los Angeles

MARGARET HILTON, *Study Director*
PATRICIA HARVEY, *Senior Project Assistant*

Acknowledgments

This report is a summary of a workshop on science education and development of 21st century skills convened by the National Research Council (NRC) Board on Science Education. The workshop would not have become a reality without the generous support of the National Institutes of Health Office of Science Education and the following members of the Partnership for 21st Century Skills: Apple, Inc.; Ford Motor Company Fund; Intel, Inc.; National Education Association; Pearson PLC; and SAP (no longer a member of the Partnership for 21st Century Skills).

We thank our colleagues who served on the planning committee, each of whom brought deep and varied expertise to the process of planning the workshop. Their diverse expertise in science teacher education, curriculum development, the role of technology in science learning, child and adolescent development, and cognitive science added greatly to the success of the endeavor. Although the planning committee played an important role in designing the workshop, the members did not participate in writing this report.

We are especially grateful to the experts who quickly responded to our request for background papers: Eric Anderman, Ohio State University, and Gale Sinatra, University of Nevada at Las Vegas; Rodger Bybee, Biological Sciences Curriculum Study (retired); Douglas Clark, Arizona State University (now at Vanderbilt University); Janet Kolodner, Georgia Institute of Technology; Joseph Kracjik, University of Michigan–Ann Arbor; Maria Araceli Ruiz-Primo, University of Colorado; Christian Schunn, University of Pittsburgh; and Mark Windschitl, University of Washington–Seattle.

We also thank the many experts, including sponsors, who participated as presenters, panelists, and discussants: Elizabeth Carvellas, NRC Teacher Advisory Council; Emily DeRocco, The Manufacturing Institute; Bruce Fuchs, National Institutes of Health Office of Science Education; Janis Houston, Personnel Decision Research Institutes–Minneapolis; Kenneth Kay, Partnership for 21st Century Skills; and Susan Koba, science consultant, Omaha.

This workshop summary has been reviewed in draft form by individuals chosen for their diverse perspectives and technical expertise, in accordance with procedures approved by the Report Review Committee of the National Research Council. The purpose of this independent review is to provide candid and critical comments that will assist the institution in making its published report as sound as possible and to ensure that the report meets institutional standards for objectivity, evidence, and responsiveness to the charge. The review comments and draft manuscript remain confidential to protect the integrity of the process. We thank the following individuals for their review of this report: Susan Albertine, LEAP States Initiatives, Association of American Colleges and Universities; Julie Bianchini, Department of Education, University of California, Santa Barbara; Norman G. Lederman, Department of Mathematics and Science Education, Illinois Institute of Technology; and Marcia C. Linn, Graduate School of Education, University of California, Berkeley.

Although the reviewers listed above provided many constructive comments and suggestions, they were not asked to endorse the content of the report, nor did they see the final draft of the report before its release. The review of this report was overseen by Jan Hustler, Partnership for Student Success in Science, San Jose State University. Appointed by the National Research Council, she was responsible for making certain that an independent examination of this report was carried out in accordance with institutional procedures and that all review comments were carefully considered. Responsibility for the final content of this report rests entirely with the author and the institution.

We are grateful for the leadership and support of Michael Feuer, executive director of the NRC Division of Behavioral and Social Sciences and Education, and Heidi Schweingruber, acting director of the Board on Science Education. We also thank Patricia Harvey, senior project assistant, for her valuable contributions to the design and implementation of the workshop agenda and the writing of this report, as well as for her flawless logistical support throughout the project.

Arthur Eisenkraft, *Chair*
Margaret Hilton, *Study Director*
Planning Committee on Exploring the Intersection of Science Education and the Development of 21st Century Skills

Contents

1	Introduction	1
2	Intersections of Science Standards and 21st Century Skills	16
3	Adolescents' Developing Capacity for 21st Century Skills	30
4	Promising Curriculum Models I	40
5	Promising Curriculum Models II	51
6	Science Teacher Readiness for Developing 21st Century Skills	61
7	Assessment of 21st Century Skills	70
8	Synthesis and Reflections	85
	References	106

Appendixes

A	Workshop Agenda and Participants	119
B	Biographical Sketches of Steering Committee Members, Presenters, Panelists, and Staff	126

1

Introduction

An emerging body of research suggests that a set of broad "21st century skills"—such as adaptability, complex communication skills, and the ability to solve nonroutine problems—are valuable across a wide range of jobs in the national economy (Levy and Murnane, 2004; National Research Council, 2008a). However, the role of K-12 education in helping students learn these skills is a subject of current debate. Some business and education groups have advocated infusing 21st century skills into the school curriculum, and several states have launched such efforts (Partnership for 21st Century Skills, 2009a; Sawchuk, 2009). Other observers argue that focusing on skills detracts attention from learning of important content knowledge (Mathews, 2009; Ravitch, 2009).

To explore these issues, the National Institutes of Health Office of Science Education and the Partnership for 21st Century Skills requested the National Research Council Board on Science Education to conduct a workshop on science education as a context for development of 21st century skills. Science is seen as a promising context because it is not only a body of accepted knowledge, but also involves processes that lead to this knowledge. Engaging students in scientific processes—including talk and argument, modeling and representation, and learning from investigations—builds science proficiency (National Research Council, 2007a). At the same time, this engagement may develop 21st century skills. For example, developing and presenting an argument based on empirical evidence, as well as posing appropriate questions about others' arguments, may develop complex communication skills and nonroutine problem-solving skills. The sponsors charged the Board on Science Education to:

Plan and conduct a public workshop to explore the intersection of science education and 21st century skills. This activity will build upon the work of a previous workshop held in May of 2007, which focused on the identification of 21st century workforce skills and the available evidence in support of that identification process.

Among the questions to guide the steering committee in their planning process are the following:

1. How much overlap is there between the 21st century skills that evidence suggests may be critical for future workforce needs and the knowledge and abilities that are the focus of current efforts to reform science education, particularly those reforms based on developmental psychology and cognitive science?

2. What are the unique domain-specific aspects of science, as well as the conventions and practices of science itself, that appear to hold promise for developing potential 21st century workforce abilities?

3. What are the promising models or approaches for teaching these abilities in science education settings? What, if any, evidence is available about the effectiveness of those models?

4. What is known about transferability of these abilities to real workplace applications? What might have to change in terms of learning experiences to achieve a reasonable level of skill transfer?

The Board on Science Education convened an expert planning committee, chaired by Arthur Eisenkraft (University of Massachusetts, Boston), to design and conduct the workshop. As a first step to meet its charge to build on the May 2007 workshop, the planning committee and staff developed preliminary definitions of five 21st century skills that emerged as important at the earlier workshop (see Box 1-1):

1. adaptability,
2. complex communication/social skills,
3. nonroutine problem-solving skills,
4. self-management/self-development, and
5. systems thinking.

Research suggests that these five skills are increasingly valuable in the workplace. Autor, Levy, and Murnane (2003), economists who studied changes over time in job tasks throughout the national economy, found that computers were eliminating tasks that involve solving routine problems or communicating straightforward information. Based on this analysis, Levy and Murnane (2004) conclude that nonroutine problem-solving skills and complex communication and social skills are increasingly valuable in the labor market. Papers prepared for the May 2007 workshop

> **BOX 1-1**
> **Preliminary Definitions of 21st Century Skills**
>
> 1. **Adaptability:** The ability and willingness to cope with uncertain, new, and rapidly changing conditions on the job, including responding effectively to emergencies or crisis situations and learning new tasks, technologies, and procedures. Adaptability also includes handling work stress; adapting to different personalities, communication styles, and cultures; and physical adaptability to various indoor or outdoor work environments (Houston, 2007; Pulakos et al., 2000).
> 2. **Complex communication/social skills:** Skills in processing and interpreting both verbal and nonverbal information from others in order to respond appropriately. A skilled communicator is able to select key pieces of a complex idea to express in words, sounds, and images, in order to build shared understanding (Levy and Murnane, 2004). Skilled communicators negotiate positive outcomes with customers, subordinates, and superiors through social perceptiveness, persuasion, negotiation, instructing, and service orientation (Peterson et al., 1999).
> 3. **Nonroutine problem solving:** A skilled problem solver uses expert thinking to examine a broad span of information, recognize patterns, and narrow the information to reach a diagnosis of the problem. Moving beyond diagnosis to a solution requires knowledge of how the information is linked conceptually and involves metacognition—the ability to reflect on whether a problem-solving strategy is working and to switch to another strategy if it is not working (Levy and Murnane, 2004). It includes creativity to generate new and innovative solutions, integrating seemingly unrelated information, and entertaining possibilities that others may miss (Houston, 2007).
> 4. **Self-management/self-development:** The ability to work remotely, in virtual teams; to work autonomously; and to be self-motivating and self-monitoring. One aspect of self-management is the willingness and ability to acquire new information and skills related to work (Houston, 2007).
> 5. **Systems thinking:** The ability to understand how an entire system works; how an action, change, or malfunction in one part of the system affects the rest of the system; adopting a "big picture" perspective on work (Houston, 2007). It includes judgment and decision making, systems analysis, and systems evaluation as well as abstract reasoning about how the different elements of a work process interact (Peterson et al., 1999).

provided evidence that these two skills and adaptability, self-management/self-development, and systems thinking are important in the rapidly growing sector of "knowledge work." For example, electrical engineers in sales often work at client sites, where they must adapt to a new work culture and apply systems thinking and complex communication skills to gain understanding of client needs and identify systems solutions tailored to meet them (Darr, 2007).

At the other end of the occupational spectrum, ethnographic and survey research indicates that low-wage service workers, such as restaurant servers, require adaptability and nonroutine problem-solving skills to meet the needs and desires of unique customers (Gatta, Boushey, and Appelbaum, 2007). Low-wage workers with higher levels of communication and problem-solving skills earn higher wages and are more likely to be promoted than those with lower levels of these skills (Maxwell, 2006). The elimination of layers of management across the economy has increased demand for individuals with self-management/self-development skills (Houston, 2007).

These definitions were provided to the background paper authors as a starting point for their reviews of various aspects of science education, with the understanding that they could modify or reinterpret the definitions as appropriate in the context of science education. The authors were allowed flexibility, because the definitions themselves are multidimensional. For example, the definition of complex communication/social skills and nonroutine problem-solving skills in Levy and Murnane (2004) is based not only on a quantitative analysis of shifts in job tasks but also on cognitive science. In their view, both of these skills involve expertise to discern patterns in a broad span of information and metacognition, the ability to reflect on whether a problem-solving or communication strategy is working, and to revise it if necessary. Similarly, the broad skill of self-management/self-development includes dimensions related to motivation and monitoring and regulating one's own learning. Although the definitions are presented in the context of the workplace, as they emerged from the May 2007 workshop, later chapters of this report indicate that dimensions of the five skills are valuable in learning generally, including science learning.

As a second step in creating a framework for the workshop, the planning committee considered how to interpret its charge, specifically the four questions that were "among the questions to guide the steering committee in their planning process." After discussion, the committee decided that the workshop would not fully address Question 4, which focuses on the transferability of 21st century skills to workplace applications.[1]

This decision reflects the fact that research on science learning has moved away from an earlier focus on domain-general reasoning strategies that may be transferable across different content areas (e.g., Inhelder and Piaget, 1958; Kuhn and Phelps, 1982). Researchers now view science learning as a complex process involving knowledge of the specific natural phenomena being studied, along with general reasoning strategies and an understanding of how scientific explanations are generated (National Research Council, 2007a). The planning committee's decision also reflects

---

[1] One of the papers does address development of argumentation skills not only in the domain of science, but also in other domains of knowledge (Clark et al., 2009).

recent research indicating that transfer is much more than simply taking knowledge and skills learned in one domain and applying them to another; instead, it is an active process that involves ongoing learning, sometimes in collaboration with others (e.g., Bransford and Schwartz, 1999). The planning committee elaborated the other questions in the charge, to arrive at a list of guiding questions for the workshop (see Box 1-2).

The planning committee used this list of questions to define topics for the background papers and structure the workshop agenda. Each workshop session was designed to address one or more of the guiding questions, and the agenda generally followed the order of the questions in this list (see Appendix A).

The structure of this report follows the structure of the workshop agenda. This chapter continues with the remarks that opened the workshop, followed by a summary of the panel discussion of demand for 21st century skills. Chapters 2 through 7 describe subsequent workshop sessions, summarizing both the presentations and the discussions. The final chapter includes a synthesis and summarizes participants' and committee members' reflections on the workshop.

BOX 1-2
Workshop Guiding Questions

1. What are the areas of overlap between 21st century skills and the skills and knowledge that are the goals of current efforts to reform science education?

2. What is the state of research on children's and adolescents' developing ability to tackle complex tasks in the context of science education?

3. What unique, domain-specific aspects and practices of science appear to hold promise for developing 21st century skills?

4. What are the promising models or approaches for teaching these skills in science education settings? What, if any, evidence is available about the effectiveness of those models?

5. How may development of 21st century skills through science education help prepare young people for lifelong learning, work, and citizenship (e.g., making personal decisions about health, making political decisions about global warming, making workplace decisions)?

6. What is known about how prepared science teachers are to help students develop 21st century skills? What new models of teacher education may support effective teaching and student learning of 21st century skills, and what evidence (if any) is available about the effectiveness of these models?

OPENING REMARKS AND PARTICIPANTS' INITIAL COMMENTS

Carlo Parravano (Merck Institute for Science Education) opened the workshop by thanking the sponsors for their desire to build educational programs on a strong research base. He explained that the workshop builds on earlier activities sponsored by the National Institutes of Health Office of Science Education, including a planning meeting on education for 21st century skills in 2005[2] and the May 2007 Workshop on Research Evidence Related to Future Skill Demands (National Research Council, 2008a).

Bruce Fuchs (National Institutes of Health Office of Science Education) observed that the 2007 workshop had engaged economists and labor market researchers in looking ahead to try to see what kinds of skill sets the national economy would require in the near future (National Research Council, 2008a). Describing his plans to support a future workshop on assessment of 21st century skills, he cautioned that the assessments used by private industry to measure these skills are too labor-intensive and expensive for use by school districts. Referring to what he sees as a backlash against 21st century skills, Fuchs contested two arguments. First, although he agreed with critics who argue that so-called 21st century creative thinking and collaboration were actually needed in Plato's time (Mathews, 2009), Fuchs suggested that this is not an important topic to argue about. Second, disputing the argument that focusing on skills leads to content-free teaching (Rotherham, 2008), Fuchs asserted, "We at the NIH want kids to have lots and lots of science content." However, he noted that this criticism does point toward important questions about which skill sets developed in science education may be domain-specific and which may be transferable to workplace problems.

Eisenkraft observed that, although he had previously been unaware of 21st century skills, he found that, as he read about demand for them (Levy and Murnane, 2004; National Research Council, 2008a), he began to see these skills everywhere. He noted that the National Science Teachers Association and the Partnership for 21st Century Skills had created a task force to develop a map of such skills in science. Eisenkraft then listed the skills used as a framework for the conference (see Box 1-1). He suggested several reasons why these skills may be uniquely important now.

In the area of communication skills, the growing diversity of the U.S. student population poses new communication challenges. Eisenkraft gave the example that earth science and physics textbooks often refer to waves on a beach, yet many students in the Boston Public Schools have never actually been to the beach. Similarly, chemistry books, when discussing the concept of balancing a chemical equation, often suggest that it is similar

---
[2]See http://www7.nationalacademies.org/cfe/21_st_Century_Skills_Planning_Meeting.html.

to baking brownies, in which one combines certain amounts of various ingredients. Most students today purchase brownies and are unfamiliar with baking, Eisenkraft said. The world of people adding and mixing measured ingredients to make brownies, he said, "is not the America we live in," yet textbook authors assume it is when they try to communicate with students. Although communication and problem-solving skills have always been important, Eisenkraft went on, society now wants everyone to have these skills, not just an educated elite. This may be another reason to focus on developing 21st century skills.

At the same time, however, Eisenkraft identified important unanswered questions, especially the question of definitions. He gave the example of widespread enthusiasm about the *National Science Education Standards* (National Research Council, 1996), which masked different understandings of the meaning of these education standards. Some textbook publishers quickly claimed that their textbooks already met all of the new standards, although, in Eisenkraft's view, the "visionary" standards document called for far-reaching changes in textbooks. Similarly, he said, a term such as "nonroutine problem solving" may mean different things to different people.

In another example of the importance of definitions, Eisenkraft said that science teachers often claim to use an "inquiry" approach to teaching. He described two quite different classrooms he observed in a recent visit to a school district. In one, the teacher lectured about the history of the theory of the atom, never mentioning that the atomic model had been revised over time on the basis of new evidence. When Eisenkraft asked why he never mentioned a model or evidence, the teacher replied that it was because he was using an inquiry approach. In the other classroom, the teacher engaged the students in a science activity and circulated around the room, but she did not talk with them or ask questions. When Eisenkraft asked why she had not used the opportunity to pose questions about the concepts being studied, she replied that it was because she was using an inquiry approach. Eisenkraft said that, while both of these teachers embrace the notion of inquiry, neither of them uses an inquiry approach as he understands it.

Eisenkraft asked whether science education may already be supporting development of 21st century skills. As a science textbook author, he said, he has never focused on such a goal, yet he wonders if he may have inadvertently incorporated them into his textbooks. He said this question would be explored at the workshop. And he asked what changes in science education might be required if 21st century skills were accepted as an important goal. Finally, he asked whether, if many young people developed these skills, this would advance the goals of science education, such as increasing scientific literacy among the public (American Association for the Advancement of Science, 1993). As an example of how science education goals might be advanced, Eisenkraft asked, "Would fewer people believe that aliens had visited the earth? Would more people believe in evolution?"

An Interactive Workshop

Eisenkraft introduced a tool for soliciting participants' ideas: carbonless notebooks that were distributed at the start of the workshop. He and other panel moderators would invite participants to write down their reflections at designated times, he said, and the planning committee would like to review their written comments in order to understand what the audience was thinking. He noted that some of the written reflections might be used without attribution in the summary report.

Eisenkraft then invited the audience to participate in a KWL process (Ogle, 1986), designed to encourage reflection and learning, in which K stands for what one knows, W for what one wants to know, and L for what one has learned. He asked participants to divide the first notebook page into three sections: "what I know, what I think I know, and what I would like to know about the intersection of 21st century skills and science education." After giving participants time to fill out these initial thoughts, Eisenkraft asked for volunteers to share what they had written. Their comments included the following:

- I think I know the meaning of the five skills, but I would like to know the components of each in more detail.
- How would classroom science teaching have to change in order to develop these skills, and what would be needed to support such change?
- My grandchildren, ages 2 and 6, already know much more than I did at those ages, demonstrating the rapid pace of change. Although we might have some idea of the skills needed in 2020, we have no clue of the skills required in 2099.
- Do parents, students, and science educators understand the critical importance of these skills in innovation and engineering?
- I know that the 21st century is different, I think I know that the government doesn't contribute very well to developing 21st century skills, and I want to learn about the policy implications of the need for these skills.
- Kids need better science knowledge to make informed decisions.
- Why has there been no discussion of the first 5 years of life, when much development takes place?
- I think that 21st century skills are important for individual success, but is science the best subject to develop them?

- Is science "enough" to support development of 21st century skills? How might the skills be integrated into other school subjects?
- Darwin's development of the theory of natural selection, based on close observation of animals, is "a great example of 21st century skills back in the 19th century." Complex scientific thinking skills have always been for a small, elite group, but what will happen when science educators try to develop them among all students?

DEMAND FOR 21ST CENTURY SKILLS

Session 1, moderated by William Bonvillian (Massachusetts Institute of Technology), focused on how the five skills manifest themselves in job performance. Bonvillian introduced the panel members: Emily DeRocco, (Manufacturing Institute of the National Association of Manufacturers and former assistant secretary of labor), Janis Houston (Personnel Decisions Research Institute), and Ken Kay (Partnership for 21st Century Skills).

DeRocco reframed the issue to focus on how the skills manifest themselves in company performance. A survey of manufacturers conducted by her association (Deloitte Development, 2005) found that 80 percent reported shortages of skilled employees across all occupations in their firms. In terms of the kinds of skills needed, the respondents most frequently cited basic employability skills, including attendance, timeliness, and work ethic; problem-solving skills; and reading, writing, and communication skills. These skill clusters are quite similar to self-management/self-development, nonroutine problem solving, and complex communication skills, respectively.

DeRocco said that manufacturers responding to a recent survey view innovation as integral to company growth, competitiveness, and shareholder value (Andrew, DeRocco, and Taylor, 2009). Survey respondents indicated that the education and skills of the workforce are the single most critical element of successful innovation, while also reporting a lack of skilled workers.

DeRocco argued that companies whose workforces lack 21st century skills are at a disadvantage in dealing with such challenges as the convergence of technology and manufacturing and the need to quickly move new products to market to beat the intense global competition. This is why, she said, manufacturers believe it is imperative to better educate the workforce not only in science, but also in 21st century skills.

Kay described a report on young people's readiness for work (Casner-Lotto and Barrington, 2006). Over 400 business executives and managers were asked to rank the relative importance of 20 skills and fields of knowledge to the job success of new workforce entrants at three education levels: high school, two-year college or technical school, and four-year college. The respondents ranked three skills among the top five most important skills and fields of knowledge for all three groups of new entrants: (1) professionalism/work ethic, (2) teamwork/collaboration, and (3) oral communication. In comparison, science knowledge was ranked 17th in importance in the list of 20 skills and fields of knowledge for high school graduates and 16th in importance for two- and four-year college graduates. When asked which skills and knowledge fields would become even more important over the following five years, critical thinking/problem solving, information technology application, teamwork/collaboration, and creativity/innovation were at the top of the list, and science knowledge was ranked 16th in growing importance.

To support the earlier point that focusing on 21st century skills need not reduce attention to content knowledge, Kay described a collaboration between the business community and the public school system in North Carolina, one of the states that has joined the Partnership for 21st Century Skills. In response to concerns about workforce preparation expressed by the state's life science industry, the North Carolina state education department engaged educational psychologist John Bransford to develop a new assessment of genetics knowledge and skills. Bransford drew on a certification exam for genetics counselors, revising both the content knowledge of genetics and genetic diseases and also the skill requirements to make them appropriate for the tenth grade level. Bransford and colleagues embedded this content and skills in a new type of assessment that challenges students to learn genetics by engaging them in playing the role of genetics counselors in different scenarios.[3] The new assessment, Kay said, "does away with this dichotomy between content and skills," because it simultaneously requires deep content knowledge and 21st century skills.

Houston responded to the question of how the five broad skills in Box 1-1 manifest themselves in job performance by first explaining her role as a corporate consultant. She and her colleagues help companies develop employment tests, establish standardized job expectations for use in performance appraisals, and define the critically important competencies[4] for successful performance across jobs in a corporation. When she reviewed her firm's recent reports in these three areas on such major companies as IBM, American Express, Microsoft, Best Buy, and Verizon, she was surprised to see how frequently the five broad skills appeared. Although these compa-

---

[3] See http://life-slc.org/?p=590.
[4] As Houston indicated, the term "competencies" refers to broad skills that are valuable across different jobs in a company. Because the five 21st century skills are also valuable across different jobs, the terms "skills" and "competencies" are equivalent, and workshop speakers used them interchangeably.

nies use somewhat different terminology, she said, the underlying skills are the same.

Houston gave some examples of the specific work behaviors that she has used as job performance standards or expectations for each of several competencies identified as critically important for the corporations she works with. She first described the "drive to achieve" competency, which includes such work behaviors/performance standards as setting and accomplishing difficult project goals, motivation to produce work of world-class quality, and staying focused and persistent in overcoming obstacles. These behaviors, she said, reflect "self-management/self-development, with a bit of adaptability." Next, she presented the competency "taking ownership," which includes taking responsibility for difficult or unpopular tasks, working effectively and productively without a lot of supervision, and accepting responsibility for one's own mistakes. Although the language is somewhat different, she said, these behaviors, too, are manifestations of the skill of self-management/self-development.

Noting that a number of companies have some kind of collaboration or communication competency that is clearly linked to complex communication skills, Houston listed the job performance expectations set for these competencies:

- Building, maintaining, and using a network of colleagues to increase productivity;
- Demonstrating a good understanding of coworkers' points of view and working with them;
- Demonstrating a thorough knowledge of the matrix structure when collaborating;
- Presenting complex technical information to nontechnical audiences in an understandable way; and
- Matching the mode of communication to the requirements of the situation.

Houston then listed several examples of job performance expectations she developed for the competency of nonroutine problem solving:

- Suggesting improvements to standard operating procedures that enhance quality and efficiency,
- Using difficult or novel problems as opportunities to innovate,
- Finding creative solutions to problems others have been unable to solve,
- Finding innovative ways of improving productivity with fewer resources, and

- Thinking beyond paradigms and methods when necessary to solve problems.

Among the five skills, Houston said, systems thinking is somewhat harder to define, but she identified some job performance standards she has developed that clearly apply to systems thinking:

- Learning and understanding how one's own work responsibility fits into the larger company's strategy, values, and goals;
- Understanding how one's own work might affect other groups and organizations; and
- Investigating issues or situations from multiple perspectives to get a more complete picture.

Bonvillian observed that the three panelists had clearly described real employer demands for talents and capabilities. Their remarks demonstrate that 21st century skills are not just an abstract theory and that this workshop has real "on the ground implications," he said. He then asked the panelists to describe what has changed in the business environment that makes these skills more important, critical, or valuable for individuals and companies than in the past.

DeRocco responded that, when she was assistant secretary of labor for employment and training, she worked with industry associations to develop models of the key competencies required in several sectors, including advanced manufacturing, energy, information technology, health care, construction, and hospitality. The core academic and workplace competencies that make up the foundation of these pyramid-shaped models (State of Minnesota, 2009) are quite similar to the five skills, including adaptability, speaking/presentation and listening skills, problem solving and decision making, and motivation. DeRocco said that, in a recent meeting to discuss these models, employer representatives were struck by the fact that the core academic and workplace competencies were exactly the same across sectors. They viewed this shared demand as an opportunity to assist workers laid off in one industry sector because of the current downturn to transition to another sector—provided the workers have the core competencies.

Returning to Bonvillian's question about change, DeRocco said the education system has become more focused on abstract, theoretical knowledge and has lost the core competencies needed in the workplace. Kay added that the flattening of organizations means that individuals lacking in self-management skills are not employable. He said he has seen teachers develop self-management skills through writing assignments that require students to set goals, assess their own progress, and improve the quality of their own work.

Kay said that the shift of employment out of manufacturing and agriculture and into the service sector, which now accounts for 86 percent of all jobs (Franklin, 2007), is another change increasing demand for 21st century skills. He described a paper presented at the May 2007 workshop about the rapidly growing number of degreed engineers who are working in sales (Darr, 2007). The paper indicates that engineers selling computer applications require communication, sales, and networking skills, along with their technical knowledge, and proposes that the most rapidly growing jobs in the U.S. labor market should be called "techno-service" occupations, because they require both technical knowledge and service-oriented communication skills (Darr, 2007).

Houston added another driver of increased demand for 21st century skills—mergers and acquisitions, which create larger, more diversified companies, requiring employees with adaptability and complex communication skills. Systems thinking is also required, as technical, sales, and service employees more often collaborate, work in teams, and share responsibility for customer satisfaction.[5]

Agreeing with DeRocco that new technology is driving increased demand for 21st century skills, Houston said that one way technology does this is by enabling remote work. At IBM, for example, she encountered individuals who had worked for the same supervisor for five years without ever meeting in person. Remote workers participating in virtual teams need communication skills to select the most appropriate mode of communication, such as knowing when to pick up the phone instead of sending e-mail. For remote workers, self-management/self-development is critical, she said.

Reiterating Kay's point that firms have restructured and eliminated layers of management, Houston observed that this has increased demand for a particular kind of problem solving. In the past, she said, companies said they needed creativity or innovation, but now they ask for "innovative solutions that work . . . within the infrastructure of the organization or that are cost-effective." Houston suggested that this growing demand reflects the reality that creative experts are no longer isolated in research and development departments but have become part of larger teams that share responsibility for developing practical solutions to business problems.

Kay added that the accelerating pace of change, the result of global competition, is another factor increasing demand for 21st century skills. Bonvillian noted that, as science and technology advance, the content will change from year to year, so the ability to master content (self-development) may become more important than knowledge of particular content. He then invited the audience to reflect on what they had learned and what they

---

[5] The increased integration of sales and technical work is discussed in Darr (2007).

wanted to know, using their notebooks. Participants' written comments are summarized in Table 1-1.

In response to Bonvillian's call to share the written reflections, a teacher shared her concerns about her students' employment prospects. She said she sees her former students working at places like discount retail stores, making such low salaries that many are forced to work two or three jobs. They do not receive any direction or management or job training, she asserted. She asked what the incentive was for students to think about 21st century skills, when the workplace does not seem to encourage the use of these skills or help to develop them.

Another participant suggested that the government ought to demand 21st century skills and asked whether it was agile enough to do so. Houston responded that her consulting work for many different federal agencies has identified demand for such skills. Eisenkraft returned to Bonvillian's point about the change in science and technology content. While agreeing that scientific knowledge changes, he observed that organizing principles allow

TABLE 1-1 Participant Responses to the Discussion of Demand for 21st Century Skills

I learned/ I think I know	The business community endorses 21st century skills.
	Industry is desperate for workers with 21st century skills.
	Business and current education policy concerns are not on same page about the importance of 21st century skills.
	Technology has changed faster than the school system can keep up.
	There is a mismatch between 21st century skills and low-wage jobs.
I want to know	How are 21st century skills defined?
	What is the value added of 21st century skills to the economy as a whole?
	Do all workers need 21st century skills?
	What kinds of jobs are out there? (A workforce with 21st century skills will not accept dead-end jobs.)
	What can engineering education bring to the table as a context for developing 21st century skills?
	How can Congress help or hurt the development of 21st century skills?
	Why isn't industry supporting public education to a greater extent?
	How can global ethics and social responsibility be addressed by 21st century skills to support development of a new business model that focuses on more than just economic growth?
Proposals	Link industry performance expectations and assessments with education.
	Convince the public that these skills are essential for students.
	Increase awareness of existing science education programs that develop 21st century skills (e.g., NASA science education, Project Tomorrow).

NOTE: The responses indicate few areas of intersection with science education.
SOURCE: Workshop participants' written comments.

fields of science to incorporate new facts. He asked whether there might be similar principles in the world of work that would guide workers in understanding the changing content and context of their jobs.

DeRocco responded that the organizing principles in the world of work are the foundational personal, academic, and workplace competencies in the models she had mentioned earlier, which provide the capacity for lifelong learning.[6] For the individual, she said, it is less important to "know everything the moment you walk in the door" than to possess these core characteristics and values.

Houston cautioned that the group should not minimize the importance of science content or job content, as both teachers and bosses remain impressed by people who know many facts. She suggested that employers view 21st century skills as an addition to core knowledge. The ability to quickly add to the knowledge base or synthesize it, she said, may be increasingly important, and this process involves 21st century skills. She explained that, in her competency modeling, she often includes a competency called "functional" or "technical" competency, in recognition of the fact that employees and teams are responsible for knowing and understanding the content of their work.

Kay asked if there was a trade-off between breadth and depth in science education. He expressed concern that requiring teachers to cover too much information would reduce time for developing critical thinking and problem solving related to particular science concepts (National Research Council, 2005, 2007a). He asked if there might be six or eight great science challenges that would require students to demonstrate both content mastery and 21st century skills and whether this might require giving up some content.

Kay then said that, although the panel had focused on the corporate perspective, he hoped they would not ignore the individual, especially at a time when so many people who have mastered the content of their jobs have been laid off. He offered two points of clarification. First, he said that the 86 percent of jobs in the service economy should not be equated with working at a discount retail store. He noted that the workshop participants are part of the service economy, which includes education, health care, scientific research, and all types of economic activity other than manufacturing and agriculture (Franklin, 2007). Second, he argued that all workers, whether employed at the high or low end of the service economy, are in danger of being laid off and will need 21st century skills in order to survive and advance in the current economy.

---

[6] See http://www.careeronestop.org/CompetencyModel/pyramid_definition.aspx.

2

Intersections of Science Standards and 21st Century Skills

This chapter focuses on two questions:

1. What are the areas of overlap between 21st century skills and the skills and knowledge that are the goals of current efforts to reform science education?
2. What are the unique domain-specific aspects and practices of science that appear to hold promise for developing 21st century skills?

To address these two questions, the planning committee commissioned a paper that assesses the extent to which the educational goals included in current science standards incorporate all or some elements of the 21st century skills that emerged from the May 2007 workshop (see Box 1-1).

ARE 21ST CENTURY SKILLS FOUND IN SCIENCE EDUCATION STANDARDS?

Christian Schunn (University of Pittsburgh) focused his examination of science standards on the *National Science Education Standards* (National Research Council, 1996) and on the standards of nine states that have joined the Partnership for 21st Century Skills: Iowa, Kansas, Maine, Massachusetts, New Jersey, North Carolina, South Dakota, West Virginia, and Wisconsin (Schunn, 2009). He thought that these states would be particularly interested in whether their standards incorporated elements of 21st century skills.

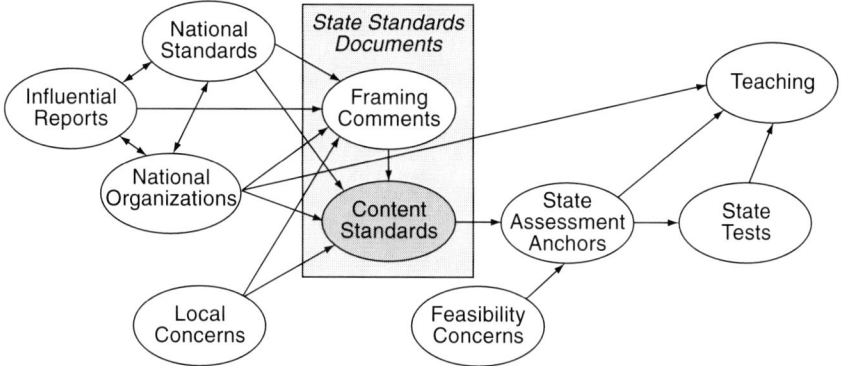

FIGURE 2-1 National, state, and local influences on science teaching.
SOURCE: Schunn (2009).

Schunn gave his rationale for focusing on science standards, rather than other elements of the education system. He said that different visions of the goals of science education, including those advanced in influential reports (e.g., National Research Council, 2005, 2007a), those included in state and national science standards, and those embodied in state science assessments, may influence science teaching and learning (see Figure 2-1). Although state science assessments have an especially strong influence, their content is changing rapidly as states respond to the science testing requirements of the No Child Left Behind Act of 2002. In contrast, state science standards change far less frequently, because creating and reaching consensus on standards is difficult and time-consuming (National Research Council, 2008b). Therefore, the analysis focusing on standards is likely to hold true for several years.

Schunn then discussed his comparison between the five skills and the *National Science Education Standards* (National Research Council, 1996). These national standards address not only student science learning, but also science teaching, professional development, assessment, and other aspects of science education; his comparison included the student learning and science teaching standards. Schunn said that the student learning standards include eight categories of goals (National Research Council, 1996, p. 6):

1. Unifying concepts and processes in science
2. Science as inquiry
3. Physical science
4. Life science

5. Earth and space science
6. Science and technology
7. Science in personal and social perspective
8. History and nature of science

Among these categories of learning goals, Schunn found science as inquiry and science and technology most relevant to 21st century skills. For example, the science as inquiry standard includes references to communication skills and to planning and selecting appropriate evidence. The science and technology category includes technological design, which involves systems thinking and nonroutine problem solving.

Because the nine sets of state standards draw on these national standards, Schunn said, he observes a "family resemblance" among them. All nine sets of standards include the first four categories listed above, a common core of content. However, Schunn found greater variability in how students are expected to engage with these content areas. Some states focus primarily on basic understanding of core theories, ideas, and facts, while other states call for students to be able to solve particular types of problems in the content area or to be able to describe patterns or explain phenomena.

Schunn noted that all of the state standards include process strands that are presented separately from the subject-matter areas, just as the national standards separate science as inquiry from the subject areas of physical science, life science, and earth and space science. These process strands present various goals for students, such as the use of appropriate scientific instrumentation, design and implementation of controlled experiments, and replication to test the validity of proposed solutions. This separation of subject-matter content from science process in state and national education standards is not supported by the research evidence, which indicates that development of science process skills is closely intertwined with—and supports—learning of science content (National Research Council, 2005, 2007a).

Schunn observed that many of the nine sets of state standards also include goals for student skills and knowledge in the design of technology, and these goals overlap extensively with 21st century skills.

Schunn then compared the nine sets of state standards with the five 21st century skills used as a framework for the workshop. He divided the broad definitions of these skills into components in order to analyze the extent to which science standards might develop each broad skill.

Schunn based his approach on cognitive research and theory indicating that skills and knowledge have components, that learning of the components occurs through practice, and that transfer can occur only when components of the new situation overlap with components of the old situation (Klahr and Carver, 1988; Singley and Anderson, 1989; Thorndike and Woodworth, 1901). Using this approach, Schunn created a five-point degree-of-overlap scale:

> 4—Strong whole skill: The skill is found almost in its entirety in the standards in a strong form likely to produce high levels of performance if the standards are met.
> 3—Weak whole skill: The skill is found almost in its entirety in the standards in a weak form, either because it is made optional or described vaguely.
> 2—Strong component skill: Only one or two components of the larger skill are found in the standards, but those elements are met to a high degree.
> 1—Weak component skill: Only one or two components of the larger skill are found in the standards, and even then only a weak form, either because they are made optional or described vaguely or are implicit in the activities of a listed standard.
> 0—None: The skill is completely absent.

Overall Level of Overlap

Overall, Schunn found a moderate level of overlap among the five broad skills, the nine state standards, and the *National Science Education Standards* (see Figure 2-2). For example, four sets of state standards and the national standards include one or two components of adaptability in a weak form (Level 1), whereas one set of state standards includes one or two components of adaptability in a strong form (Level 2), and four sets of standards do not include adaptability at all. Similarly, four states and the national standards include a few components of complex communication/social skills in a weak form (Level 1), and seven states and the national standards include a few components of nonroutine problem solving in a weak form (Level 1). Seven states either include only one or two components of self-management/self-development in a weak form (Level 1) or do not include any components of this skill at all. Only for systems thinking is the degree of overlap higher. As shown in Figure 2-2, four states include most components of systems thinking but in a weak form (Level 3); two states and the *National Science Education Standards* include a few components of the skill in a stronger form (Level 2); and three states include a few components of the skill in a weak form (Level 1).

20 INTERSECTION OF SCIENCE EDUCATION AND 21ST CENTURY SKILLS

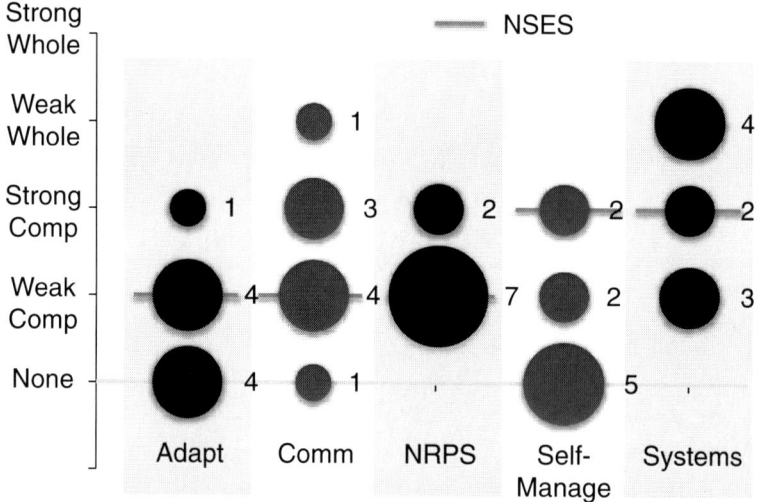

FIGURE 2-2 Frequency count of the degree of overlap of state standards with each of the five 21st century skills.
KEY: Adapt = adaptability; Comm = complex communication/social skills; NRPS = nonroutine problem solving; Self-Manage = self-management/self-development; Systems = systems thinking.
NOTE: The horizontal lines indicate the rough level found in the *National Science Education Standards* (NSES).
SOURCE: Schunn (2009).

Adaptability

Schunn divided the definition of adaptability into the following four components (Pulakos et al., 2000):

1. Ability and willingness to cope with uncertain, new, and rapidly changing conditions on the job;
2. Handling work stress;
3. Adapting to different personalities, communication styles, and cultures; and
4. Physical adaptability to various indoor or outdoor work environments.

He first considered the extent to which the nine sets of state standards include goals related to "ability and willingness to cope with uncertain, new, and rapidly changing conditions on the job." He observed that several sets of state standards call for students to engage in design of technological processes, and that design involves identifying new problems as they emerge and developing appropriate solutions, which is similar to this component of adaptability. Although most states do not connect design skills with adaptability in the student's own life and career, West Virginia calls for students to be able to "investigate, compare and design scientific and technological solutions to personal and societal problems."

Schunn said he did not find the second component of adaptability, "handling work stress," in any of the sets of state standards. The third component of adaptability, "adapting to different personalities, communication styles, and cultures," appears in the *National Science Education Standards*' Teaching Standard "E," which refers to supporting collaboration (National Research Council, 1996). Schunn explained that, because today's public school population is very diverse, supporting collaboration should indirectly lead to learning to adapt to different personalities, communication styles, and cultures. He noted that West Virginia's science standards were the most explicitly related to this component of adaptability (West Virginia Department of Education, 2006, pp. 11, 16, 21):

> Demonstrate the ability to listen to, be tolerant of, and evaluate the impact of different points of view on health, population, resources and environmental practices while working in collaborative groups.

Finally, Schunn said that he did not find the final component of adaptability, "physical adaptability to various indoor or outdoor work environments," in any of the state or national science standards.

Complex Communication/Social Skills

Schunn divided complex communication/social skills into five components. He found that these skills appear more strongly than adaptability in science standards. The majority of the nine states and the national standards refer to communication of scientific findings orally and in writing. However, the standards emphasize the cognitive, rational aspects of communication, rather than the social ones.

The first component of this skill, "select key pieces of a complex idea to express in words, sounds, and images, in order to build shared understanding" (Levy and Murnane, 2004), appears in a few state standards. For example, the Wisconsin standards explicitly refer to trying to build understanding, and the Kansas inquiry standards include detailed goals for written and oral communication, including constructing arguments and responding appropriately to critical comments. Turning to the second component of this skill, "social perceptiveness," Schunn said some standards refer to attending to the views of others, but none directly refers to social perceptiveness.

The third component of this skill is "persuasion and negotiation." Schunn said that the concept of persuasion appears often in the science standards' call for students to use evidence to support a scientific argument. He noted that the Kansas inquiry standards for grades 8-12 are especially detailed in this area, when describing the expectation that the student "actively engages in communicating and defending the design, results, and conclusion of his or her investigation" (Kansas Department of Education, 2007, p. 63). This standard includes the following components (p. 63):

a. Writes procedures, expresses concepts, reviews information, summarizes data, and uses language appropriately.
b. Develops diagrams and charts to summarize and analyze data.
c. Presents information clearly and logically, both orally and writing.
d. Constructs reasoned arguments.
e. Responds appropriately to critical comments.

Schunn observed that the social elements of persuasion and negotiation are not mentioned in the science standards. Turning to "instructing," the fourth component, Schunn said that the emphasis on clear communication and explanation in most of the state standards is relevant to instructing others. However, effective instruction involves not only clear communication, but also actively assessing the knowledge of others, and the latter aspect of instruction is not mentioned in any of the science standards.

The fifth component, "service orientation," did not emerge in any of the state or national science standards Schunn reviewed.

Nonroutine Problem Solving

Schunn divided this skill into six components:

1. Narrow the information to reach a diagnosis of the problem.
2. Ability to reflect on whether a problem-solving strategy is working and switch to another strategy if the current strategy is not working.
3. Creativity to generate new and innovative solutions.
4. Integrating seemingly unrelated information.
5. Recognize patterns not noticed by novices.
6. Knowledge of how the information is linked conceptually.

Across all of the science standards, Schunn found nonroutine problem solving at a relatively low level on his five-point degree-of-overlap scale. Although the science standards include some components of this skill by calling for students to be engaged in inquiry and technological design, these two types of learning activities may be scripted and routine, so that they do not support development of nonroutine problem solving.[1]

Schunn found that science standards frequently referred to the first component of nonroutine problem solving—the ability to narrow information in order to reach a diagnosis of the problem (Levy and Murnane, 2004)—and they did so in a variety of ways. For example, the Iowa inquiry standards for grades 9-12 include the benchmark "reads and interprets scientific information." Within this benchmark is a standard for grade 10 students to "select best evidence" (Iowa Area Education Agencies, 2005). The New Jersey inquiry standards for grade 4 state: "Develop strategies and skills for information-gathering and problem-solving, using appropriate tools and technologies" (New Jersey Department of Education, 2004, p. E-5). North Carolina science and technology standards call for grade 8 students to be able to "identify problems appropriate for technological design; develop criteria for evaluating the problem or solution; identify constraints that must be taken into consideration" (North Carolina Public Schools, 2004, p. 83).

The second component is the "ability to reflect on whether a problem-solving strategy is working and switch to another strategy if necessary." Schunn observed that revision of strategies is somewhat similar to revision of theories, which is mentioned in the national standards and in several sets of state standards. In addition, a few state standards discuss this component in their technological design standards, because redesigning a product or process involves moving beyond an existing solution and deciding that a new approach is required. For example, the Maine inquiry and technological design standards for grades 9-12 state: "Students use a systematic process, tools and techniques, and a variety of materials to design and produce a solution or product that meets new needs or improves existing designs" (Maine Department of Education, 2007, p. 87).

Turning to the third component, "creativity to generate new and innovative solutions," Schunn said that most states mention creating new scientific theories and/or new designs. For example, West Virginia standards call for students in grades 5 through 7 to "apply skepticism, careful methods, logical reasoning and creativity in investigating the observable universe" (West Virginia Department of Education, 2006, pp. 24, 29, 34).

---

[1] A recent study of high school science laboratories found that students' laboratory experiences are typically scripted and routine (National Research Council, 2005).

The Wisconsin Standard G (Science Applications) calls for students at grade 8 to "propose a design (or re-design) or an applied science model or machine that will have an impact in the community or elsewhere in the world" (Wisconsin Department of Public Instruction, 2008, Standard G.8.4). With regard to the fourth component, "integrating seemingly unrelated information," Schunn observed that, although the science standards do not refer to this specifically, both science and design involve integration of various kinds of information.

According to Schunn, the fifth and sixth components of nonroutine problem solving, "recognize patterns not noticed by novices" and "knowledge of how the information is linked conceptually," develop naturally through extensive practice and growing expertise in a domain (Chase and Simon, 1973; Chi and Koeske, 1983; Gobet and Simon, 1996). He cautioned that, because these skills are intertwined with deep domain knowledge, they may not readily transfer to another domain, such as from science education to the workplace. Noting that both analysis of patterns and knowledge of how information is linked conceptually are core aspects of scientific reasoning, he said he found these components in some of the state science standards.

Self-Management/Self-Development

Schunn divided the broad skill of self-management/self-development into six components:

1. Ability to work remotely, in virtual teams.
2. Ability to work autonomously.
3. Self-motivation.
4. Self-monitoring.
5. Willingness and ability to acquire new information related to work.
6. Willingness and ability to acquire new skills related to work.

Overall, he found a high degree of overlap between these six components and the teaching standards of the *National Science Education Standards* (National Research Council, 1996), but much less overlap with state science standards.

Although a few sets of state standards and the national standards mention the "ability to work collaboratively," which is related to work in virtual teams, the standards do not explicitly discuss virtual collaboration. Similarly, "self-motivation" is not directly discussed in any of the sets of science standards. However, students may develop self-motivation through the *National Science Education Standards*' Teaching Standard E, which calls on teachers to "enable students to have a significant voice and decision about the content and context of their work and require students to take responsibility for the learning of all members of the community" (National Research Council, 1996, p. 46).

Schunn said that the goal of developing self-monitoring in students is reflected in the *National Science Education Standards'* Teaching Standard C, which directs teachers to "guide students in self-assessment" (National Research Council, 1996, p. 42). He also found aspects of self-monitoring in two sets of state science standards, including the following element of the Kansas science inquiry standards for grades 8-12: "Evaluates personal preconceptions and biases with respect to his/her conclusions" (Kansas State Department of Education, 2007, p. 63).

Schunn said that, because gathering new information to inform scientific theories is related to the component "willingness and ability to acquire new information related to work," all of the sets of science standards reflect this aspect of self-development/self-management. With regard to "willingness and ability to acquire new skills related to work," Schunn said that, although the science standards do not explicitly mention this component, it could be developed through Teaching Standard E of the *National Science Education Standards*, discussed above.

Systems Thinking

Turning to systems thinking, Schunn divided this broad skill into two components: "systems analysis" and "systems decision making." Noting that systems analysis is "what scientists do," he said that references to this component appear in all nine sets of state standards. For example, the Massachusetts technology/engineering standards call for students in grades 6-8 to be able to "identify and describe three subsystems of a transportation vehicle or device, i.e., structural, propulsion, guidance, suspension, control, and support" (Massachusetts Department of Education, 2006, p. 89). In contrast, systems decision making, which he described as "the bread and butter of engineering," appears less frequently in the sets of science standards included in the review.

Conclusion

Schunn concluded that he sees the current state of science standards, in relation to 21st century skills, as "the glass half full." If all students learned exactly what the national and state science standards call for, he said, they would not possess all components of the five broad 21st century skills. Nevertheless, the components of the skills they had learned would provide a foundation for further learning of these complex skills over time, including through in-depth training at work. Schunn cautioned against cursory comparisons of textbooks, standards, or other materials to assess the degree to which they include 21st century skills. It is easy to find a particular phrase, such as "communication skills," on one page of a textbook and conclude that it is aligned, but this will not ensure that students develop the skills and are able to transfer them to new contexts. He also warned that including 21st century skills in state science standards does not necessarily lead to increased student learning of science. Despite his finding that state and national standards are "half full" of 21st century skills, student scores on recent state science assessments show that they are very weak in science knowledge and skills.

Schunn said that, on the basis of his finding that engineering design standards call for development of more components of the five skills than do science standards alone, it would be valuable for states to include engineering design standards in their sets of science standards. He also predicted that assessing student learning of the five skills would be very difficult. He predicted that later workshop presentations would show that engaging students in large, team-based design projects supports development of 21st century skills, because of what such projects require (Kolodner, 2009; Kracjik and Sutherland, 2009): working in a team would develop communications and social skills, a large team size requires adaptability and self-management skills, and the design process requires problem-solving and systems thinking skills.

RESPONSE

Struck by the wide variation in the different states' science standards, Bruce Fuchs said he was both pleased and surprised by the extent to which 21st century skills were included. He sounded two notes of caution about the analysis. First, he observed that science standards generally overestimate the quality of actual classroom lessons. Second, he noted that Schunn's analysis probably underestimates the level of development of 21st century skills in a few exemplary lessons.

Fuchs said it was important to define such terms as "nonroutine problem solving" in order to understand how best to teach and assess them. For example, he noted that, in *Teaching the New Basic Skills*, Murnane and Levy (1996) argue that a critical skill for obtaining a middle-class job is the ability to solve a semistructured problem by creating and testing a hypothesis. They provide detailed examples of this type of problem solving at work on an auto assembly line and in a life insurance company. Fuchs said this raised the question in his mind of whether nonroutine problem solving, as defined by Levy and Murnane (2004) and used in the workshop, includes hypothesis-testing. If so, he said, science standards might include this type of problem solving more frequently than Schunn found in his analysis. He added that problem solving and the other 21st century skills could be taught in history or other subjects, not only in science.

Fuchs then asked Schunn whether 21st century skills are developed only in specific domains of knowledge, or whether they might be transferred to other domains, such as from a science classroom to a workplace problem. He noted that Klahr and Nigam (2004) found that, with appropriate instruction, 77 percent of a class of third and fourth graders learned how to design a controlled experiment. These young students were also able, at a later time, to transfer this new skill, applying it to evaluate other students' poster presentations of their experiments at a science fair. Fuchs described these findings as "amazing," because, in his own experience, not all graduate students develop the skill of control of variables. Those who do master it, he said, could apply it not only in science, but also in business and other domains.

Schunn responded that, although much of the research in the learning sciences focuses on transfer of knowledge and skills, researchers rarely obtain evidence of transfer. He also noted that Klahr and colleagues found that helping students master control of variables was more difficult and time-consuming in urban schools than in more affluent, suburban schools (Li, Klahr, and Siler, 2006). He went on to say that he realized how complex systems thinking is when he asked some faculty colleagues to review his paper on development of systems thinking among K-12 students. Some of the reviewers said that, although they think about how concepts relate to each other in their particular fields, such as earth systems and biological systems, they do not consistently employ systems thinking.

DISCUSSION

Moderator Marcia Linn (University of California, Berkeley) invited the audience members to talk with their neighbors about their reflections on the session and to write down their questions for the presenters. Participants raised several questions, including:

1. Do science teachers possess 21st century skills?
2. What is the relationship between standards and actual teaching practices?
3. Are schools culturally ready for new approaches to teaching 21st century skills that might appear chaotic when compared with teaching practices in other classrooms?
4. What 21st century skills should students learn in school as a complement to informal learning outside school and as a base for deepening their skills in higher education and the workplace?

Schunn responded that these questions shared a focus on systems and recognition that systems-level thinking is essential in order to reform science education. He agreed with an idea embodied in the fourth question: Teaching and learning of 21st century skills in formal elementary and secondary education is only one component of a larger system of lifelong informal and formal learning. He called for thinking about how to reform that larger learning system but cautioned that he and other workshop participants were ill prepared to do so, because their U.S. education had not developed their skills in systems thinking.

Linn added that science standards are driving decisions in schools today that do not support development of 21st century learning skills. Current science standards, she said, push teachers and schools to cover many topics superficially, reducing students' interest in science and discouraging teachers from leading inquiry activities that would develop their ability to think deeply about science.

Responding to the question about whether teachers possess 21st century skills, Fuchs said that teachers solve nonroutine problems, engage in complex communication, and work in teams. However, he said, he was discouraged by the findings of an evaluation of implementation of curriculum supplements developed by the National Institutes of Health Office of Science Education. He noted that his office collaborated with well-known curriculum development organizations to create "really exemplary inquiry-based materials" on life sciences topics, such as genetics and bioethics. However, when evaluators observed classrooms using these materials, they found teachers emphasizing memorization and vocabulary, rather than inquiry. He concluded that teacher readiness is a big problem, even if good instructional materials are available.

Linn agreed with Fuchs that it is important to think not only about the kinds of instructional materials that support learning of 21st century skills, but also about how to support both teachers and students in developing these skills. She said that she often visits classrooms using materials that she and her colleagues have developed, and that, because the materials are delivered in online environments, they require students to conduct experiments and reflect on their learning (Linn, Davis, and Bell, 2004). Teachers can read student reflections immediately, in order to find out whether they understand a particular lesson or concept. The challenge, Linn said, lies in helping teachers respond appropriately if they find out students are not learning. She described a video in which a teacher responds to students' lack of understanding of aspects of evolution by delivering a lecture, noting that the teacher lacked the repertoire of skills needed to teach more effectively.

Schunn added that there is a continuum of levels in systems thinking. Individuals do not simply either possess or lack this kind of thinking, but instead may build from a very rudimentary awareness that elements of a system affect each other to a much more sophisticated understanding of how concepts or elements interact within systems. Considering education as a system, Schunn said that feedback is "a completely broken construct." The focus of feedback is on assessments to help teachers monitor student progress, with no attention to developing teachers' skills to change the course of instruction in response to the feedback.

3

Adolescents' Developing Capacity for 21st Century Skills

This chapter discusses adolescents' cognitive and social development and the role of high-quality science instruction in fostering their development of 21st century skills. A commissioned paper on this topic explores one of the workshop guiding questions: What is the state of research on children's and adolescents' developing ability to tackle complex tasks in the context of science education? The chapter summarizes the paper, the response, and the ensuing discussion. It also summarizes small-group discussions of the first three workshop sessions.

TEACHING AND LEARNING ABOUT SCIENCE IN THE 21ST CENTURY

Educational psychologists Eric Anderman (Ohio State University) and Gale Sinatra (University of Nevada, Las Vegas) discussed adolescents' cognitive abilities related to the five 21st century skills, highlighting approaches science educators can use to create social learning contexts that foster these skills (Anderman and Sinatra, 2009). Anderman began by saying that he and Sinatra hoped to convince the audience of the importance of helping science teachers understand how adolescents learn, what skills they possess and lack, and what makes them unique. He said they also hoped to convince the audience that, if high school students have bad experiences in science classrooms, this will turn them off from advancing in science studies and from entering science careers.

Adolescents' emerging cognitive abilities present unique challenges and opportunities for science educators, Anderman said. However, secondary science teachers, who often have a strong background in a science, such as biology or chemistry, may not have an equally strong background in adolescent development. As a result, teachers may be unsure of what motivates their students and how they engage in scientific inquiry. At the same time, the depth and breadth of science classes expand at the high school level, offering students greater opportunities to build on their elementary and secondary science knowledge, to enroll in multiple courses, and to take specialized classes, such as anatomy and environmental science.

Adaptability

Sinatra explained that the good news from the research is that adolescents have the capacity to think and reason adaptively about science. However, this ability must be fostered and supported by teachers, peers, and learning environments. Even if teachers provide the required levels of support, many high school students lack the base of rich, interconnected science knowledge that is necessary for adaptive reasoning. Students' lack of content knowledge is partly due to the weakness of current science curriculum materials, which often aim to introduce many different science topics, rather than treating a few concepts in depth (Vogel, 2007).

Sinatra said that adaptability requires not only a rich knowledge base, but also the willingness to engage in effortful thinking and to consider alternative points of view or to engage in scientific argumentation. Some students are low in what social psychologists call "need for cognition;" that is, they do not necessarily seek or enjoy opportunities to engage in effortful thinking (Cacioppo et al., 1996). Students also vary in their degree of openness to new ideas and in their beliefs about the nature of knowledge, and these factors influence the likelihood that a given student will experience a change in his or her knowledge base and be willing to engage in scientific argumentation.

Sinatra said that developing adaptable thinking in science requires that students are willing to have their ideas publicly challenged, although such challenges can be psychologically uncomfortable during adolescence, when young people are very sensitive to the perceptions of their peer groups. In some cases, a challenge to one's point of view can even be seen as a threat to identity. For example, if students identify themselves as belonging to a group that believes in creationism, it may be difficult to learn about evolution. These social and psychological concerns can lead students to avoid adaptive thinking about scientific concepts and processes. Sinatra said that the hallmarks of adaptability include both recognizing the need to change one's thinking and also the willingness to change it, based on one's view of scientific knowledge as subject to change on the basis of new evidence.

Complex Communication

Anderman observed that communication is critical in science, as scientific investigations are increasingly conducted by team members who must communicate clearly and effectively with each other. While arguing that adolescents are capable of communicating effectively about abstract concepts, he cautioned against the assumption that they will "naturally" learn communication skills. Written communication in science is a complex psychological process, he said, requiring self-regulation (recognizing one's own strengths and weaknesses as a learner and applying effective learning strategies); construction of complex sentences; adopting a scientific style; and self-confidence both in science and as a writer. As a result, science teachers must be well prepared in order to help students develop skills in science writing.

Effective science teachers incorporate techniques into their instruction that facilitate the development of oral communication skills. One useful cooperative learning technique is referred to in the literature as "jigsaw" (Slavin, 1995). In jigsaw, each member of a group is responsible for becoming an expert in a particular area. That expert then reports back and teaches the other members of the group about the specific topic. In this manner, students scaffold and support each other's communication as they learn the necessary information.

Nonroutine Problem Solving

Sinatra noted that, because most scientific problems worth solving are ill structured, they require nonroutine problem solving or what is often referred to as "thinking outside the box." Successful problem solving, she said, requires a strong base of relevant knowledge and both "the skill and the will" (Paris, Lipson, and Wixson, 1983). Although many adolescents have the skill (the reasoning, metacognitive, and self-regulatory skills necessary to solve science problems), fewer have the will (the motivation to approach difficult problems and to persist toward a solution).

Sinatra explained that science instruction can be designed to support these skills, motives, and dispositions by providing practice in solving problems connected to student interests. She cautioned against engaging students in "overly simplistic inquiry tasks," referring to Arthur Eisenkraft's earlier comments about the very different ways that science teachers understand inquiry (see Chapter 1). If students develop a sense that science is not very complex, she said, this could reduce their motivation to approach and persist in solving nonroutine science problems.

Self-Management

Anderman observed that there is a large body of research literature on what psychologists call self-regulation and what most people would call self-control. The literature includes studies focusing on how adolescents learn to control, regulate, and monitor their use of various learning strategies (Zimmerman, 2000). As with adaptability and complex communication skills, the research indicates not only that adolescents are capable of self-management in their learning, but also that this skill does not develop naturally, in the absence of teaching and coaching. One element of self-management, Anderman said, is self-efficacy, or the confidence that one has the capacity to engage in and complete a learning task (Schunk and Zimmerman, 2008). Confidence is extremely important for self-management of learning in science subjects (Greene and Azevedo, 2007).

Anderman explained that research has begun to identify instructional strategies that build students' self-management in learning. Teachers can model self-regulated learning strategies (Pintrich, 2000; Zimmerman, 2000, 2001) and provide students with some autonomy. They can encourage students to evaluate the quality of their own work, providing opportunities for students to go back and correct any earlier errors (Pintrich and Schunk, 2002; Schunk and Ertmer, 1999).

Systems Thinking

Sinatra began by paraphrasing Chi's (2005) definition of complex systems as "systems with multiple component parts and processes that interact in ways that give rise to emergent phenomena." For example, Sinatra said, consider the V-shaped pattern of birds in flight. The pattern is the result of each bird seeking the path of least resistance; it is not predictable from examining the flight mechanics of individual birds but can be understood only from examining the interaction of the birds' individual actions (Chi, 2005). Addressing complex scientific problems, such as predicting the effects of a pandemic, forecasting tornados, or understanding the decline in the bee population, requires systems thinking.

Although the research indicates that adolescents have the capacity for systems thinking, it also illuminates the difficulty they have in understanding emergent systems, even when they are given specific, directed instruction. One promising approach to supporting student understanding of systems is the use of computer simulations (Jackson et al., 1996).

Creating Adaptive Motivational Contexts in Science Classrooms

Anderman said that science teachers' small daily decisions can increase students' motivation to develop 21st century skills in the context of science learning. The tasks they give to students, the ways they group students, the amount of choice and types of rewards they provide, and the expectations they have for students all have profound effects on motivation (Anderman and Anderman, 2009). Teachers' everyday discourse in science classrooms also influences whether students view the goal of science class primarily in terms of performance—i.e., getting a good grade—or in terms of really learning and mastering science (Maehr and Midgley, 1996). Students whose goal is mastery are more likely to enroll in further science classes. The bottom line, he said, is that students who have bad experiences in science classes are likely to rule out science as a career.

Teachers also influence student goals through assessment practices. The types of assessments given are "completely predictive of cognitive engagement," Anderman said. Assessments focused on sorting out students based on ability lead to problems, including increased cheating (Anderman et al., 1998) and students' placing a lower value on science (Anderman et al., 2001). Tests that stress memorization of facts lead to less cognitive engagement than tests that are focused on solving real-world problems and that build upon prior knowledge (Dole and Sinatra, 1998). When teachers focus on the external motivation of tests, they often decrease intrinsic motivation, he said.

Sinatra concluded the presentation by offering a list of recommendations to strengthen development of 21st century skills in high school science:

1. Foster productive learning environments;
2. Promote active engagement based on connections to students' personal interests and career goals;
3. Develop requisite knowledge, skills, and dispositions necessary for science literacy and to support nascent science career choices;
4. Capitalize on learning progressions by revisiting earlier content in more depth;
5. Promote an inquiry and problem-based learning approach to science instruction;
6. Use assessments that focus on higher order learning;[1] and
7. Provide professional development for secondary science inservice and preservice teachers that includes adolescent development and motivation.

---

[1] The term "higher order learning" refers to Bloom's (1956) taxonomy of learning objectives. In this taxonomy, the lower levels (or orders) include recall, comprehension, and application of information, and the higher levels (or orders) include analysis, synthesis, and evaluation.

Commenting on the list, Sinatra observed that connecting to students' personal interests is especially important to motivate adolescents. The third recommendation—building knowledge and positive dispositions toward science learning—is critical, given that today's students, the "Google generation," are accustomed to instantly accessing vast amounts of information. The fourth recommendation, she said, reflects the fact that high school provides an opportunity to build on earlier learning, on adolescent students' growing cognitive capabilities, and on teachers' expertise, which is generally greater than at the elementary school level. Next, promoting inquiry and problem-based reasoning are promising methods to capitalize on adolescents' growing adaptability, complex communication, and other 21st century skills. Finally, she urged that professional development for secondary science teachers include information about adolescent development and motivation.

Respondent Susan Koba (science education consultant) asked Anderman and Sinatra about implementing these suggestions: How can the goals be made accessible for teachers, especially in school districts that aren't engaged in partnerships with colleges and universities, and so lack support and access to the most recent research?

Anderman responded that administrators' support was the most important factor in implementing the suggested improvements in teaching practice. For schools and districts not near higher education institutions, he said, there is good research-based information on science teaching available on the Internet. He acknowledged that it is a challenge to translate that information into changes in classroom teaching.

Sinatra added that all seven of their ideas for improvement are related to teacher professional development. She expressed the view that long-term, sustained, supported teacher professional development is the key to implementing the suggestions, adding that she would like to see a greater emphasis on professionalization of teachers.

DISCUSSION

In response to questions, Anderman said that, if the principal supports innovation and supports teachers in taking risks and trying different approaches to instruction, then teachers will change their teaching practices. He said that teachers in other countries have more time to collaborate in lesson planning and reflection. In some countries, he said, teachers spend only 35 percent of their day with the students, allowing time to talk with one another and learn together. In the United States, team teaching is often implemented in order to provide time for such collaborative learning, but in reality, he said, this does not yield the time needed to support teachers in planning and reflecting on their lessons.

Koba added that, in schools in which administrators support innovation, teachers form learning communities; they study cases of teaching practice and examine student work to develop their skills. Observing that good video case studies are available, she said the problem is the lack of consistent support systems and access to the best research. Sinatra added that technology can assist in teacher professional development. Although teachers are sometimes isolated in their classrooms, they often have computers with Internet access, allowing them to communicate with scientists and other science teachers and to obtain real-time support. Sinatra described a classroom at the University of Nevada, Las Vegas, that is equipped with multiple microphones and cameras, so that preservice teachers can observe everything, from the teacher leading the class to an individual student's paper. After watching short periods of instruction, the students can stop and discuss their observations. In addition, the teacher in the classroom can contact the students who are observing to ask their opinions about an activity or segment of instruction that has just taken place.

In response to a question about undergraduate science instruction, Sinatra noted that biology professors are not often given instruction in teaching and may not have much prior teaching experience, so they may not always be good role models for how to teach. Anderman added that the university pulls faculty members in different directions and does not reward them for devoting extra time to improving their teaching, trying more creative approaches, or observing their colleagues' classes.

Kenneth Kay asked about terminology. He noted that "self-management/self-development" is different from the Partnership for 21st Century Skills' term, "self-direction," and that Anderman and Sinatra had introduced yet another term, "self-regulation." He asked if it was possible to agree on a shared system of naming this and other 21st century skills in a way that would be understandable to students, parents, and employers. Sinatra replied that it was unlikely that everyone would agree on a common term, because different disciplines have their own histories of developing and defining terms. However, she said that it was possible to improve understanding of the common elements underlying the different terms.

Anderman added that it takes 10 to 20 years for people in the single field of educational psychology to agree on how to define a term, although the problem of focusing on terminology isn't unique to this field. He agreed with Kay that it is very important to communicate with parents in language they can understand, for example, by avoiding terms like "self-regulation." The challenge, he said, is to ensure that when talking with parents about these skills, everyone is talking about the same concepts.

In response to a question about stages of development of 21st century skills, Anderman said that their paper focused on adolescents as requested.

He cautioned that trying to develop these skills only in grades 9 through 12 would not lead to the outcomes desired by employers. Instead, he said, development of these skills should begin at age 5 or even younger and should continue throughout elementary, middle, and high school. Such a process of continuous development would require major changes in science teaching, he noted.

Sinatra cautioned against thinking that 21st century skills are something an individual either has entirely or lacks completely. Elementary students may begin to develop the rudiments of these skills, and adolescents can develop them more fully, a first-year teacher in a master's program will develop them even more fully, and an expert teacher may possess even higher levels of the skills.

REPORTS FROM DISCUSSION GROUPS

Workshop participants were divided into small groups of 8-12 people to discuss what they had learned and what they still wanted to know about the topics addressed in the three preceding sessions.

Moderator William Sandoval (University of California, Los Angeles) invited each reporter to briefly summarize the group's report. Susan Albertine (American Association of Colleges and Universities) said that, through a rich discussion, her group learned that they need to take a systems perspective on development of 21st century skills from prekindergarten through graduate school. Following an initial lack of clarity about how to define such a systems perspective, they agreed that they were focusing on developing young people's skills over a long trajectory of years in all curriculum areas, not only science. Albertine said the group agreed about the importance of 21st century skills in broad terms, but people began to disagree when they moved into specific questions, such as the difference between self-management and self-regulation. This group would like to learn how to bridge the gap between classroom realities and the research and policy on 21st century skills and how to move beyond discussion toward creating systemic change.

Douglas Oliver (American Association for the Advancement of Science and the National Science Foundation) reported that his group learned that children and youth "probably have a greater capacity for developing 21st century skills than we older folks have." The group also learned that business support for 21st century skills could be helpful in leading reform of science education and that demand for these skills creates demand for higher level teaching skills and more independent students. The group would like to know how 21st century skills may promote education reform that has been discussed for two decades. Group members also asked how policy makers can influence implementation of 21st century skills in the school curriculum.

Ines Cifuentes (American Geophysical Union) said her group learned that business is very clear about its demand for 21st century skills, although some business representatives prefer the term "competencies" rather than "skills." At the same time, however, the group learned that the five skills used as a framework for the workshop are not well understood or agreed on. Finally, because they see 21st century skills as important for the nation, they learned that it is important to identify leverage points at which they can have an impact on developing these skills. Cifuentes also raised several questions. First, the group would like to know if science is uniquely positioned to develop these skills and, if so, whether learning these skills would actually support students' learning of science. Second, she noted that the group was unclear about what problem 21st century skills might solve. If everyone had these five skills, she asked, what problem would be solved?

Gina Schatteman (National Institutes of Health) reported that her group learned that science is an excellent vehicle for developing 21st century skills, although they believed there was no strong evidence for this view. The group also learned that discussions like this should include practitioners, after realizing that many group members were former teachers but none was currently teaching. Third, they learned that, in order to develop students' capacities for 21st century skills, it is important to sustain their natural wonder and interest. This group would like to know how to redesign educational standards for depth and relevance and how to design metrics for 21st century skills. The group suggested a longitudinal study to assess whether students were truly learning 21st century skills and later transferring them to the workplace. Schatteman said that, like other groups, hers would like to know how to apply a systems approach to analyzing and implementing development of 21st century skills.

Sinatra reported that her group learned that, because children's activities today are often highly structured by parents, they have fewer opportunities to develop self-management or self-regulation skills. Still, young people have many technology skills that they use to learn through social networking on the Internet, and this could be an asset in developing 21st century skills. The group also thought that environmental concerns, such as climate change, could motivate students to engage in deep science learning. Sinatra also reported several questions from the group:

- What will emerge from the thinking of today's students, which is very different from the thinking of previous generations?
- What level of science knowledge and skill is necessary for the general population?

- Should there be a more explicit connection between understanding of the nature of science and development of 21st century skills?
- How will adolescents' use of technology to access and share information affect their view of intellectual property?

Brian Jones (JBS International) said participants in his group were surprised to learn about the report indicating that employers view science knowledge as less valuable than 21st century skills (Casner-Lotto and Barrington, 2006). The group also learned that these skills may be difficult to assess. Finally, participants learned about the importance of self-development and systems thinking. Jones reported that the group would like to know about equity issues that should be addressed in the teaching of 21st century skills. Participants noted the focus on adolescents' unique capabilities and constraints in learning and asked what other groups—defined by gender, age, race, language skills, and/or socioeconomic status—might have special capabilities and constraints that should be considered when teaching 21st century skills.

Reflecting on the group reports, Sandoval observed that many of the groups discussed the importance of systems thinking, based on their view of education as a complex system that is difficult to understand. He described the groups' calls to identify key stakeholders or leverage points, in order to drive change, as very important, asking which facets or components of the education system are most amenable to change. Surveying the room, Sandoval noted that, although representatives of many components of the education system were present, there were still "some real gaps in who is here and who is not here." He observed that the attendees did not reflect the diversity of the United States, leading him to pose questions about who "owns" 21st century skills and whose purposes the skills might serve, if they were widely acquired.

A second theme Sandoval observed frequently in the reports was uncertainty about how to operationally define the five skills, so that they can be easily recognized and so that student learning of the skills can be measured by appropriate assessments. Arguing that it is very important to determine how to assess these skills, he posed the rhetorical questions, "How do we do that? What's a smart way of doing that?"

A final common theme in the reports, Sandoval said, was that, although the workshop focused on science education, this is not the only school subject in which these skills can be learned and practiced.

4

Promising Curriculum Models I

This chapter and the next focus primarily on two questions addressed at the workshop:

1. What are the promising models or approaches for teaching these abilities in science education settings? What, if any, evidence is available about the effectiveness of those models?
2. What are the unique, domain-specific aspects of science that appear to support development of 21st century skills?

Four papers prepared for the workshop describe promising curriculum models. In order to ensure that the papers would address both of these questions and to increase uniformity across papers, the workshop planning committee provided a set of guiding questions to each author:

1. Curriculum goals: To what extent does the curriculum model target the five 21st century skills emerging from the May 2007 workshop or similar skills defined in the context of science education for instruction?
2. Alignment with learning research: To what extent does the curriculum treat 21st century skills and content knowledge as separate or intertwined? To what extent does the model reflect research on children and adolescents' learning and development in science?
3. Assessment and evidence: Where has the model been implemented? Does the model incorporate assessment of 21st century skills? What evidence is available about development of one or more 21st

century skills among students and/or teachers engaged with the model?
4. Effectiveness and implications: What does the available evidence indicate about the impact of the model on development of 21st century skills among diverse groups of science learners? What does the evidence indicate about unique, domain-specific aspects of science that may support development of 21st century skills? Does the available evidence point to principles of instructional design for development of 21st century skills that may be applicable to other science curricula and/or teaching strategies?

This chapter summarizes the two papers presented on the first day of the workshop, and Chapter 5 summarizes the two papers presented on the second day. Chapter 8 synthesizes the evidence of intersections between science education and 21st century skills from all four papers.

ONLINE LEARNING ENVIRONMENTS FOR ARGUMENTATION

Douglas Clark (Arizona State University) presented an overview of his team's paper, which considers how engaging students in argumentation in online environments can help promote the development of 21st century skills (Clark et al., 2009). First, he explained that the team focused on argumentation, because inquiry and argumentation are at the heart of current efforts to help all students develop scientific literacy (American Association for the Advancement of Science, 1993; National Research Council, 1996). Scientific literacy, he said, involves understanding how knowledge is generated, justified, and evaluated by scientists and how to use such knowledge to engage in inquiry in ways that reflect the practices of the scientific community. Engaging students in argumentation can build this understanding and application of science processes.

Clark used the term "scientific argumentation" to describe a process in which students learn, whether in the domain of science or in another domain, to:

- develop, warrant, and communicate a persuasive argument in terms of the processes and criteria valued in science and
- construct, critique, and communicate sound and valid arguments in terms of the connections between and among the evidence and theoretical ideas.

He proposed that both of these facets of scientific argumentation are central 21st century skills. However, he cautioned, developing scientific argumentation can be quite challenging for students (e.g., Abell, Anderson, and Chezem, 2000; Bell and Linn, 2000; Kuhn and Reiser, 2005; McNeill and Krajcik, 2008a; Ohlsson, 1992; Sandoval, 2003).

To address these challenges, education researchers have focused over the past 15 years on developing computer-enhanced environments to support students in constructing arguments and engaging with one another in argumentation. Clark briefly described four examples of such learning environments. Although only one of the environments—the Web-based Inquiry Science Environment (WISE) Seeded Discussions—was developed specifically to support argumentation in the domain of K-12 science, the other three develop argumentation that is similar to scientific argumentation in terms of argumentation structure, what counts as evidence, and goals.

The Computer-supported Argumentation Supported by Scripts-experimental Implementation System (CASSIS) environment has been used by undergraduate education psychology students in Germany to collaboratively solve problem cases related to attribution theory. CASSIS uses a similar approach to that developed by Guzdial and Turns (2000) in their CaMILE online learning environment, which has been used to support collaborative learning in science and in other domains. The Virtual Collaborative Research Institute (VCRI) learning environment has been used to support argumentation and collaborative learning among secondary students in the Netherlands in the domains of history, Dutch, and social studies. The Dialogical Reasoning Educational Web Tool (DREW) learning environment has been used to develop argumentation in the domain of science policy among secondary and undergraduate students in Finland.

WISE Seeded Discussions

WISE Seeded Discussions environment focuses on grouping students together with others who have expressed differing perspectives or stances. WISE Seeded Discussions first engages students in exploring the phenomenon to be discussed through probe-based labs and virtual simulations and then supports them in constructing an explanation for the phenomenon. In order to help students focus on the salient issues and articulate clear stances, they are given drop-down menus to construct their explanation from sentence fragments identified through research on students' alternative conceptions. Once the students have submitted their explanations, they are organized into discussion groups with other students who have created explanations conceptually different from one another. Students participate in asynchronous online discussion of their explanations, in which they are encouraged to propose, support, critique, evaluate, and revise ideas. Finally, they reflect on how their ideas have changed through the discussion.

WISE Seeded Discussions have been implemented in a broad range of public middle and high school science classrooms, with data generally collected on three to six classes of students for each study. Assessments initially focused on analyses of the structural quality of argumentation among students and later expanded to investigate the conceptual and grounds quality, in addition to the structural quality, of students' argumentation (Clark and Sampson, 2005, 2007, 2008). The rubrics used in these assessments incorporate elements of complex communication/social skills and nonroutine problem-solving skills. More recent studies have used pretests and posttests to analyze gains in content knowledge.

Computer-Supported Argumentation Supported by Scripts-Experimental Implementation System

CASSIS is designed to facilitate argumentation in asynchronous on-line discussions through collaboration scripts, which specify and sequence collaborative learning activities. Developed at the University of Munich, CASSIS engages groups of three students in analyzing problem cases using a specific theory. Usually, the group's task is to first analyze three problem cases and then develop a joint solution for each case, collaboratively constructing an argument. An asynchronous, text-based discussion board is built into the environment so group members can communicate with each other as they work, and different collaboration scripts are implemented to promote and support productive collaboration among the students. For example, a script for the construction of single arguments consists of three text boxes that require students to input a claim, grounds, and qualifications as they construct the argument.

As an experimental learning environment, CASSIS has not yet been fully integrated into the core curriculum of an entire course. However, several hundred students enrolled in an educational psychology class at the University of Munich have participated in experimental sessions using CASSIS that take the place of a three-hour lecture on attribution theory—a theory of how people explain their successes and failures. Assessments of student learning through the use of CASSIS have focused on the quality of collaborative argumentation, based on analysis of transcripts of individual contributions to the online discussion. Some of this assessment has been automated.

Virtual Collaborative Research Institute

VCRI is a groupware program developed in the Netherlands, designed to support collaborative learning on inquiry tasks and research projects, allowing students to communicate with each other, access information sources, and coauthor texts and essays. While working with VCRI, students share tools designed to support the collaborative inquiry process over the course of approximately eight lessons. They start by investigating a topic, using a sources tool. Students are able to discuss the information found in these sources with other group members, using the synchronous chat tool. Students use the debate tool to help them examine and explore the arguments contained in these information sources. The debate tool enables the students to collaboratively create an argumentative map, a visual representation of the arguments in a single source or across sources. Once the argumentative maps are complete, students can transfer the lines of reasoning to the co-writer tool, a text processor that allows simultaneous editing by multiple users, to write a final report using the lines of reasoning identified and highlighted with the debate tool as a guide.

Dialogical Reasoning Educational Web Tool

The DREW environment, developed at the University of Jyväkylä, Finland, consists of several different tools designed to support collaborative activities, including a chat environment, a collaborative writing tool, and an argument diagram tool. The argument diagram tool enables users, either individually or collaboratively through a shared screen from different workstations, to construct argument boxes that include claims, arguments, and counterarguments. The boxes are connected with each other by arrows indicating whether the content of the box supports or criticizes the content of the box to which the arrow points. A completed diagram depicts the argumentative structure of a text or discussion by indicating the main thesis of the materials and showing how the thesis is supported and criticized by illustrating other arguments and counterarguments and their interconnections.

Researchers studied use of the DREW environment among secondary students working in dyads engaged in chat discussions about current science policy issues (e.g., genetically modified organisms, nuclear power, and animal experimentation). They also studied secondary school students' use of the DREW diagram tool to organize and structure arguments and counterarguments gathered from different Internet sources to be used in a joint essay-writing task. In a current study, university students are using DREW to analyze the content of scientific articles by creating argument diagrams using the DREW diagram tool.

Summary and Implications

All of the online learning environments, Clark said, are designed to implement and test instructional design principles developed through research on argumentation and the learning sciences (e.g., National Research Council, 2000). Although the environments did not specifically target the five 21st century skills as learning goals, these skills are deeply intertwined with the development of scientific argumentation. The environments thus focus on skills, habits of mind, and communication processes that are central to both science and the development of 21st century skills.

The research team's review of the research on student learning in these four environments indicates that they can support development of 21st century skills (Clark et al., 2009), Clark observed. He discussed examples related to each of the five skills, noting that they are not equally supported by the environments. Overall, the learning environments support the development of complex communication skills most strongly, followed by problem solving, self-monitoring, adaptability, and systems thinking (see Chapter 8 for a summary of the research).

Finally, Clark noted that the research has implications for the design of other science curricula and teaching strategies. He said that the four online learning environments are organized as scripts and activity structures that orchestrate and structure students' interactions with each other and the environments. Current research on these learning environments focuses on the efficacy of various configurations and structures of the scripts. These scripts and activity structures could easily be incorporated into other online and offline curricula. He concluded that the research provides evidence that a broad range of collaborative learning skills can be supported by online environments for the development of scientific argumentation.

THE BIOLOGICAL SCIENCES CURRICULUM STUDY 5E MODEL

Rodger Bybee, former director of the Biological Sciences Curriculum Study (BSCS), opened his presentation by noting that, although education policy makers and practitioners have agreed for over a decade on the goal of developing scientific literacy among all students (American Association for the Advancement of Science, 1993; National Research Council, 1996), U.S. students' scores on international comparative assessments of science show little evidence of progress toward this goal (Lemke et al., 2004).

The current concerns in the business community about the education and skills required for work are not new, Bybee went on, citing reports from the 1980s and 1990s (National Research Council, 1984; U.S. Department of Labor, 1991). More recently, he said, researchers have identified skills required for the workplace (Levy and Murnane, 2004; Murnane and Levy, 1996) that are somewhat similar to widely accepted goals for reform of science education (American Association for the Advancement of Science, 1993; National Research Council, 1996). Bybee noted that reports about workforce skill demands propose somewhat different definitions of the exact types or levels of skills required. For example, although most call for development of broad, transferable skills similar to those identified in the May 2007 workshop (see Box 1-1), others focus on more specific knowledge and skills in the fields of science, technology, engineering, and mathematics.

Now, Bybee said, educators face the challenge of clarifying the skills that are needed for work and moving from broad statements of purpose to more specific discussions of educational practice. This challenge, in turn, led to the analysis in his paper of the intersection between 21st century skills and the 5E model (Bybee, 2009).

Description of the 5E Model

Bybee explained that the current 5E model has its origins in one of several science curriculum study groups established by the National Science Foundation in the 1960s after the Soviet Union succeeded in launching the Sputnik satellite. The Science Curriculum Improvement Study (Atkin and Karplus, 1962) developed the learning cycle model, including the three phases of explore, invent, and discover (Karplus and Thier, 1967).

During the late 1980s, BSCS convened a group of experts to review and revise the learning cycle model. The group added a new first stage—engage—with the goal of increasing students' interest and motivation. And in response to teachers' requests for an approach to assess student learning using the model, the group added a final stage—evaluate. The group also changed the name of the second stage in the learning cycle from invent to explain, and the third stage from discover to elaborate. These changes led to the current 5E model: (1) engage, (2) explore, (3) explain, (4) elaborate, and (5) evaluate.

In the 5E model, the curriculum is designed to allow students to explore scientific phenomena and their own ideas. Students are invited to explain their ideas, and explanations can also be provided by the teacher or a textbook or through the use of technology. The curriculum then helps students to clarify the key concepts targeted for instruction by engaging them in new situations in which they can elaborate and extend their learning. Finally, the curriculum invites both students and teachers to evaluate the learning that took place.

Bybee explained that the model reflects research on how students learn. Rather than simply requiring students to progress through a series of exercises that are sequenced to cover certain science topics within a certain number of days, the model aims to expose them to major concepts as they arise naturally in problem situations. The model calls for structuring activities in these problem situations so that students are able to explore, explain, extend, and evaluate their own progress. The model is based on findings from cognitive research that ideas are best introduced when students see a need or a reason for their use. Seeing relevant uses of the knowledge helps students to derive meaning from the activities (National Research Council, 1999, p. 127).

Bybee said that, although it was developed in the 1980s, the 5E model also reflects more recent research, such as the research reviewed in the National Research Council study (2005) of high school science laboratories. That study concluded that laboratory experiences are more likely to support student science learning when they are integrated with other forms of instruction (National Research Council, 2005, p. 82). The 5E model represents one approach to integrating different forms of science instruction.

Implementation of the Model

In 2006, BSCS was funded by the National Institutes of Health Office of Science Education to assemble and analyze all of the available research on the 5E instructional model. Bybee acknowledged that, although BSCS had been using the model in curriculum development and implementation for 20 years, the organization had conducted little research on its effectiveness. On the few occasions he did submit proposals to conduct research studies, he said, they were rejected because funders viewed research by the developers of the model as self-serving.

The review of the available research found that the model was implemented widely across the United States and in other countries. For example, a simple Google search of the term "BSCS 5E instructional model" returns about a quarter of a million citations. Most frequently, the term appears in:

- documents that frame larger pieces of work, such as curriculum frameworks, assessment guidelines, and course outlines;
- curriculum materials of various lengths and sizes; and
- adaptations for teacher professional development, informal education settings, and disciplines other than science.

At the same time, the team found that there was not very much research available on learning outcomes among students exposed to the 5E model. Nevertheless, some studies suggest that the instructional model is more effective than alternative approaches at helping students master science subject matter (e.g., Akar, 2005; Coulson, 2002). One of the key findings in the research relates to how faithfully teachers follow the model. Students whose teachers taught with medium or high levels of fidelity to it exhibited learning gains that were nearly double those of students whose teachers did not use the model or used it with low levels of fidelity (Coulson, 2002; Taylor, Van Scotter, and Coulson, 2007).

TABLE 4-1 Evidence of Development of 21st Century Skills Through the 5E Instructional Model

Goal of 21st Century Skill	BSCS 5E Instructional Model
Adaptability	Inadequate evidence
Complex communication	Some evidence based on argumentation
Nonroutine problem solving	Strong evidence based on scientific reasoning
Self-management/self-development	Strong evidence based on attitudes toward and interest in science
Systems thinking	Strong evidence based on mastery of scientific knowledge

SOURCE: Bybee (2009).

Linking the Model to 21st Century Skills

Bybee said that his review of the available research did not find any cases in which 21st century skills were specifically targeted as desired learning outcomes in curricula based on the 5E model. The model is aimed at developing students' mastery of science subject matter, not at development of skills. With these caveats in mind, he then discussed the available evidence on development of each of the five skills (see Table 4-1). He found no evidence of development of adaptability, some evidence of development of complex communication/social skills (i.e., argumentation), and stronger evidence of development of three other skills: (1) nonroutine problem solving (i.e., scientific reasoning); (2) self-management/self-development (i.e., interest in science and science learning); and (3) systems thinking (i.e., mastery of content knowledge about complex scientific systems) (see Chapter 8).

Conclusion

Bybee concluded with four observations. First, he said that further clarification is needed about what each 21st century skill looks like and how to teach it, in order to provide concrete, explicit guidance to teachers. Second, it is important to identify specific learning outcomes associated with 21st century skills, because the preliminary definitions of these skills used as a framework for the workshop include a mixture of cognitive abilities, social skills, personal attitudes, motivational interests, and conceptual understanding. He suggested a need to tease out these different aspects in order to establish more specific learning outcomes. Third, he called for increased clarity in curriculum goals related to science education and 21st century skills. Finally, he suggested that, because the 5E model is a known quantity in the world of science education, it could serve as an excellent vehicle for developing 21st century skills, if adapted to focus on these skills.

DISCUSSION

Following both presentations, moderator Arthur Eisenkraft asked whether the environments for scientific argumentation described by Clark and the 5E model described by Bybee are intended for use in creating effective science lessons and if they reflect research on student learning. Bybee responded that the 5E model is based on Piaget's theory of how children learn at different stages of cognitive development. He tells curriculum developers to focus first on how to evaluate student progress toward learning goals and then work backward from the goals to develop the curriculum.

Eisenkraft asked whether students and teachers are aware of the underlying learning models in curricula based on the 5E model or in the online learning environments. Clark responded that increased student awareness of their own internal learning process is an important part of the scaffolding built into the online environments. If the students are still able to engage in scientific argumentation when the scaffolding is faded, he said, this indicates that they are becoming aware of their own developing skills in argumentation. Bybee said that it is important to sort out some specific learning outcomes from the list of 21st century skills to use as a starting point and then design the curriculum to help students attain these outcomes. One important outcome might be development of durable skills that are transferable to new situations, such as skills in the control of variables developed by science students in a study by Klahr and Nigam (2004).

Marcia Linn asked how to take advantage of the embedded assessments, the tasks that students are actually doing in these more complicated curriculum materials, and Clark responded that a potential advantage of the online environments is that computers continually monitor student behavior and provide some real-time assessment. Eisenkraft asked about the possibilities for wide dissemination and implementation of the two instructional models. Clark responded that the online models for scientific argumentation highlighted in his presentation are not implemented on a wide scale, but other curricula using technology, like Investigating and Questioning our World through Science and Technology (IQWST, see Chapter 5), are being used in a fairly large number of school districts. In addition, he said, the Technology Enhanced Learning in Science Center, funded by the National Science Foundation, which engages students in Internet environments for science education, has grown to include 7 school districts and over 100 teachers.[1] He predicted that other states might soon join the center.

Clark also said that WISE provides an activity authoring system that can be used by others. In addition, Yael Kali has developed a database of design principles that are useful in constructing science learning environ-

---

[1] See http://telscenter.org/about/index.html.

ments (Kali, Linn, and Roseman, 2008). Teachers, he said, could take those design principles and create a new online science curriculum unit to meet their specific needs. However, Clark reported, more often a teacher will use these design principles or the WISE authoring system to customize existing curriculum materials to meet her or his local needs.

Eisenkraft asked what specific characteristics of the 5E model would promote 21st century skills, and Bybee responded that the primary goal of the model has been to increase student mastery of subject matter, including science concepts. He said that he has learned that, in order to reach any instructional goal, it is important to focus on that specific goal. Therefore, in order to develop 21st century skills through the 5E model, these skills would have to be explicitly targeted for instruction.

5

Promising Curriculum Models II

This chapter continues the focus on promising new models for developing 21st century skills through science education begun in the previous chapter. Research conducted on both of the curriculum models described in this chapter provides support for a key premise of the workshop—that deep understanding of science content and development of skills in science processes (e.g., designing an investigation, formulating a scientific explanation based on evidence) are closely connected and mutually reinforcing (National Research Council, 2007a). The science process skills developed by students exposed to the models include facets of the five 21st century skills listed in Box 1-1.

LEARNING BY DESIGN

Janet Kolodner (Georgia Institute of Technology) explained that Learning by Design (LBD) is a project-based inquiry approach to learning science and scientific reasoning in the context of design challenges, developed by her research group in the late 1990s and early 2000s (Kolodner, 2009). Middle school students work in small groups on design challenges that require targeted science, scientific reasoning, collaboration, communication, and planning.[1] The curriculum includes a set of units that cover a half-year each of earth and physical sciences. The design challenge is one that can be achieved in the physical or virtual world, for example, designing and building a vehicle that can navigate a certain terrain (to learn about motion

---

[1] See http://www.cc.gatech.edu/projects/lbd/home.html.

and forces) or designing and modeling an erosion control system (to learn about the earth's ground processes).

Although the design challenge creates an authentic need for learning the targeted science content and skills, Kolodner emphasized that design challenges are not required in order to support the learning of complex skills. She said she is currently applying the overall approach, in collaboration with others, to develop middle school science curriculum units as part of the Project-Based Inquiry Science project.[2]

Learning Goals

Kolodner outlined several learning goals targeted in the LBD curriculum. First, the team hoped to develop students' knowledge and skills in design, including understanding a project design challenge, planning and managing time, aiming for shared solutions with understanding, developing specifications and criteria, managing trade-offs, understanding and working with real-world constraints, and gaining experience in the iterative process of design. Second, the curriculum was designed to develop knowledge and skills in science practices, including identifying what needs to be investigated and carrying out an investigation well. One of the science practices targeted for development is informed decision making, which includes reporting on and justifying conclusions and judging the trustworthiness of experimental results in order to use evidence appropriately to inform decisions. In addition, the LBD curriculum aims to help students learn to develop and articulate scientific explanations.

Kolodner said that enhancing learners' collaboration skills is another important goal of the LBD curriculum. Components of this broad goal include supporting the development of teamwork, collaboration across teams, and giving credit to individuals and teams. Finally, the curriculum aims to develop science content knowledge consistent with middle school objectives. Kolodner explained that the curriculum was originally developed with a focus on technology education, but it was revised to place greater emphasis on science content knowledge, in order to align with state and national science standards.

Alignment with Research on Learning

The LBD curriculum is based on a constructivist model of learning called case-based reasoning (Kolodner, 1993; Schank, 1982, 1999), which suggests several principles for promoting learning of complex skills, such as the five 21st century skills. First, the model suggests that learning is

---

[2]See http://www.its-about-time.com/pbis/pbis.html.

enhanced when learners have goals that they want to pursue, because they will reflect on their progress towards achieving those goals and seek explanations when their progress is not as expected. Second, it suggests that learners should have experiences that allow them to try out targeted skills in the context of working towards their goals, analyze whether they are achieving them through those skills, identify what they need to do better, and have the opportunity to try again. Third, it suggests that learners need multiple opportunities to try out each of the skills they are learning. Fourth, the model suggests that, in order to track their progress toward their goals, learners need to be able to easily identify the effects of what they are doing. Fifth, because identifying these effects may be difficult, the model suggests that learners be helped to analyze feedback, identify what they are doing well and not as well, and generate ideas about how to perform more productively. Finally, so that skills are learned in a way that is durable and transferable to new situations, learners should practice the targeted skills in varied contexts that are representative of the kinds of situations they will encounter outside the formal learning environment.

Kolodner noted that these principles for learning complex skills are consistent with the cognitive literature on skills learning and transfer (National Research Council, 2000) and on learning through shared reflection on practice (Lave and Wenger, 1991; Wenger, 1998).

In addition to reflecting the research on case-based reasoning, the LBD design was informed by problem-based learning (Barrows, 1985; Koschmann et al., 1994), an approach that has been used and studied extensively in medical schools. Problem-based learning integrates coaching, scaffolding, and reflection into learners' problem-solving experiences in order to develop both targeted content and the reasoning needed to solve problems.

Cycles of Design and Investigation

All of these bodies of literature informed the creation of the LBD curriculum, designed to encourage learning from experience. Kolodner presented an illustration of the two related cycles of experiences incorporated in the curriculum (see Figure 5-1).

The first iterative cycle involves design and redesign. It begins with "understanding the challenge," or knowing what needs to be accomplished. An understanding of the design challenge leads students to develop ideas for what they "need to do" and "need to know" in order to meet the challenge. These ideas lead the learners into the second cycle, "investigate-explore," in which they obtain and share information that they need to know to inform construction and testing of the design. If the test indicates that the design does work well enough, this leads to new ideas about what the students need to do and know, and the cycles continue.

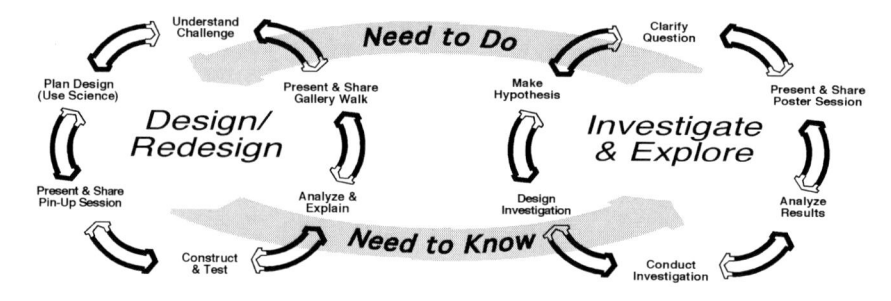

FIGURE 5-1 Design and investigation cycles in Learning by Design.
SOURCE: Kolodner (2009).

Development of 21st Century Skills

Kolodner said that comparisons of learning among students using LBD and matched classes of science students (in terms of science achievement level, teacher understanding of the material, and socioeconomic status) indicate that the curriculum develops all five of the skills used as a framework for the workshop (Kolodner, Gray, and Fasse, 2003). She argued that students learn adaptability as they work on multiple large challenges throughout the year and over several years and join different small groups. They learn complex communication skills and social interactions through presentations and discussing and writing down their ideas on a project white board. Because the challenges are complex, students need each other's ideas in small groups, and each group needs the results of the other groups, further developing communication and social skills.

Nonroutine problem solving develops through engagement with the curriculum, Kolodner said, because the students work on a variety of problems that may have many good answers. Self-management is a key goal. Students are supported in learning self-management through "launcher units," which gradually introduce the practices of scientists and engineers in contexts in which the value and purpose of these practices are clear. Self-management is also enhanced through scaffolding that helps learners be successful in engaging in design/redesign and investigation activities and by emphasizing practices and reflecting on them. Finally, students develop systems thinking in the course of working to achieve design challenges that require systems thinking, with appropriate supports.

Conclusion

Reflecting on lessons learned in the design and testing of this curriculum model, Kolodner emphasized several features that support learning of complex skills. First, emphasizing practices pushes learners to think about how they are doing things, what works and what does not. Second, introducing science practices through the launcher units helps students understand the importance of these practices while also providing an initial opportunity to use and learn about them. These introductory units promote creation of a learning community and a positive classroom culture. Third, repeated public presentations help learners develop internal scripts (Schank and Abelson, 1977) that make both presenting and listening and asking questions feel automatic. These public presentations help learners reflect on and improve their approaches to investigations and design/redesign, in a context of authentic need. Overall, she said, the curriculum promotes a shift in the roles of teachers and students toward increased initiative by the students.

INVESTIGATING AND QUESTIONING OUR WORLD THROUGH SCIENCE AND TECHNOLOGY

Joseph Krajcik (University of Michigan) opened the presentation of his paper (Krajcik and Sutherland, 2009) by explaining that Investigating and Questioning our World through Science and Technology (IQWST) is a large, multi-institution curriculum research and development project under way since 2000. The primary goal is to help middle school students develop integrated understanding of core ideas of science through coherent curriculum materials (Shwartz et al., 2008). He noted that the project team has implemented and studied elements of the curriculum in both affluent suburban schools and inner-city schools.

Project Goals

Krajcik outlined the goals of the IQWST project:

- Design, develop, and assess the next generation of middle school science materials;
- Enable teachers to effectively teach students with a variety of backgrounds;
- Explore core ideas from each scientific discipline each year; and
- Support students in building sophisticated and systematic understanding of scientific ideas and practices.

The key features of IQWST include coherence, development of curriculum driven by learning goals, a focus on learning big ideas of science over time, and project-based learning.

Krajcik said that curriculum coherence is valuable, because research has shown that it leads to integrated understanding in learners (Kali, Linn, and Roseman, 2008; Linn and Eylon, 2006; Schmidt, Wang, and McKnight, 2005). The developers of IQWST aim for coherence in learning goals, by selecting key goals that build on each other. They also strive for coherence within each 8-10-week project-based curriculum unit, coordinating among content learning goals, scientific practices, and curricular activities (Krajcik and Blumenfeld, 2006). In addition, the curriculum developers seek coherence across the separate units, coordinating the units to support how big ideas in science connect with each other.

IQWST curriculum materials are built around the big ideas of science that can help students understand the natural world—such as the particulate nature of matter. The focus on big ideas includes development of both science content and scientific practices and allows designers to revisit ideas throughout the curriculum so that student understanding becomes progressively more refined, developed, and elaborated across different science disciplines.

Learning Performances: A Key Feature of IQWST

Krajcik described the IQWST approach to integrating goals for content and skills through "learning performances." He noted that science standards often call for students to "know" or "understand" a science concept, while also including separate goals for science process skills. In IQWST, however, learning performances describe not only what it means for a learner to understand a concept, but also how the student should apply the concept, using scientific reasoning and other skills. Learning performances are based on the research team's view that students cannot learn science content without practice, and they cannot learn science practice without content.

Krajcik presented an example learning performance that integrates the content and practice standards for grades 5-8 that are included in the *National Science Education Standards* (National Research Council, 1996). The content standard focuses on understanding that the properties of a substance are independent of the amount of the sample, and the practice standard focuses on using evidence to develop explanations. The performance standards call on students to construct a scientific explanation that includes a claim, evidence, and reasoning to support the concept that different substances have different properties.

Development of 21st Century Skills

Krajcik observed that the focus of IQWST on learning performances and scientific practices supports the development of adaptability, although he did not find evidence of this (see Chapter 8). IQWST also supports development of complex communication skills, Krajcik said, noting that research on IQWST provides evidence of improvement in students' construction of written explanations. Krajcik argued that three features of IQWST—its emphasis on coherence, its focus on big ideas in science, and its emphasis on constructing explanatory models—support students in solving nonroutine problems (see Chapter 8). He added that two features of IQWST—its inclusion of scientific practices and its focus on evaluating and revising models—support growth in self-development. Reflecting on one's understanding is critical to self-development. In IQWST, learners need to consider if the model they have constructed accurately represents the phenomenon being studied. If not, then learners need to revise their models to more accurately represent it. There is evidence that students' evaluation and revision of models improves during and across units.

In addition, monitoring an investigation is critical to self-management and self-development. For example, in the eighth grade chemistry unit, *How do I get the energy to do things?*, students design and carry out a long-term experiment in which they have to study one variable that influences plant growth. They manage data collection over a 5-week period, requiring them to monitor their investigations several times a week. Krajcik repeated that IQWST is built around big ideas in science, and that tracking how one part of a scientific system affects the rest of it is a critical aspect of developing systems thinking. For example, students in sixth grade biology track the flow of energy in an ecosystem; students in seventh grade chemistry consider the mass changes in closed and open systems; and students in eighth grade chemistry investigate how matter and energy move between organisms.

Conclusion

Krajcik concluded that various studies support the claim that middle school students can learn both scientific concepts and practices through engagement with the IQWST curriculum materials. He noted that, if constructing scientific explanations, building and revising models, designing investigations, and building products reflect understanding of the five 21st century skills, then there is evidence that students can learn these skills. The greatest challenge is encouraging students to use the reasoning component when constructing scientific explanations.

Krajcik cautioned that, although the IQWST coherent curriculum materials have the potential to produce a populace that is scientifically literate and prepared for the new skill demands of the 21st century, this hypothesis requires further empirical support. He noted that, although the national field test currently under way will provide some data, the IQWST materials need to be tested using a more careful experimental design in which teachers are randomly assigned to using the IQWST materials or some other materials, and then teachers using the IQWST materials must be tracked to see if they are implementing the unit according to the designers' intent. Observing that teacher implementation of materials affects student learning (McNeill and Krajcik, 2008b), he said it is important to provide intense professional development to help teachers use the materials as intended.

In closing, Krajcik thanked his colleagues in the IQWST project, including Brian Reiser, Northwestern University; David Fortus, Weizmann Institute of Science in Israel; his coauthor LeeAnn Sutherland of the University of Michigan; and many graduate student contributors. He thanked the teachers who have been willing to test the materials in their classrooms, and National Science Foundation Program Officer Gerhard Salinger for his support of the project through the years.

DISCUSSION

Following the two presentations, moderator Carlo Parravano invited the audience members to write down their reflections and discuss them with a neighbor. After a few minutes, he called for questions. The first questioner asked how to bring these promising models into classrooms, given the data showing their effectiveness. Kolodner responded that the LBD approach has been incorporated into a comprehensive middle school curriculum called Project-Based Inquiry Science (PBIS). The new curriculum includes units lasting 8 to 10 weeks that are carefully sequenced, so the teacher is not required to decide when to introduce topics or learning activities. She added that publication of the materials has encouraged more teachers to use them.[3]

This exchange led to a discussion about adoption of published curriculum materials. Krajcik said that the IQWST materials are being used in the Lubbock, Texas, school district, which has launched an initiative on writing across the curriculum. The IQWST focus on writing scientific explanations fits well with this initiative. He noted that, although it is difficult to respond to critics who say that the IQWST materials do not meet all of the Texas science standards for grades 6 through 8, it is valuable to emphasize the broader goals of the materials, such as helping students learn to construct arguments and write scientifically. The teachers and principals in the district who are using the IQWST materials are very enthusiastic about them, he said. They are delivering presentations about the curriculum materials and promoting their use to parents and community leaders. Kolodner said that

---

[3] See http://www.its-about-time.com/pbis/pbis.html.

the same process is happening among principals and teachers who have adopted the PBIS materials.

Douglas Clark asked how the approaches to assessment used in LBD and IQWST could inform design of large-scale assessment of science process skills and 21st century skills. He observed that widely used multiple-choice examinations cost only about ten cents per student to administer, whereas the Organisation for Economic Co-operation and Development's Program of International Student Assessment costs about $42 per student to administer.

Kolodner said she was not sure that the approaches to performance assessment used by her research team could be scaled up, because they require an entire class period. She explained that, in these assessments, students are asked to solve a novel problem, working in small groups. The research team makes video and audio recordings of the discussion of the problem and also asks students to write down how they would design an investigation, gather data, and formulate an explanation to solve the problem. The team analyzes the recordings and written documents to assess individual and group performance.

Krajcik said that an outside evaluator is conducting a study that tracks the performance of IQWST students and a matched comparison group of similar students over three years, from sixth through eighth grade. The evaluator originally designed multiple-choice assessments matched to the benchmarks for these grades included in the *National Science Education Standards*. For four years, Krajcik said, the IQWST advisory board expressed its disagreement with this assessment instrument. In the fifth year, the evaluator finally agreed to include items in the assessment focusing on the use of evidence and the construction and revision of explanatory models. He noted that it is challenging to create such test items, but they will be included in the assessments administered to the IQWST group and the comparison group when the two groups are in the seventh and eighth grades. Krajcik noted that the IQWST team has just begun collaborating with some of the best assessment experts in the nation to develop new approaches to tracking development over time in students' understanding of how matter interacts and changes.

In response to a question about integrating learning of science content knowledge and skills with learning of other subjects, Kolodner said that she has observed this in some schools. As a practical matter, she said, when researchers or curriculum developers are creating science curriculum materials, they cannot assume that the science teacher will use the materials in collaboration with teachers of other subjects. Krajcik said that many teachers using IQWST in different parts of the country connect the curriculum's focus on explanations with other subjects. These teachers, he argued, recognize that argumentation can be useful in English and history classes.

A participant observed that the LBD curriculum model emphasizes students' need to know certain science concepts and processes and asked Krajcik whether the need to know also plays a role in the IQWST materials. Krajcik responded that each IQWST curriculum unit begins with a large driving question that builds not only coherence of learning activities but also students' motivation. Typically, he said, the unit begins by engaging students in activities related to a phenomenon in order to see the importance of the driving question. For example, in one unit, the teacher asks students to close their eyes and then releases an odor into the classroom. In this unit, the students return several times to this opening encounter with the phenomenon of odor, building models to explain why something that is a source can reach their noses. Through this process, they gain understanding of the particulate nature of matter and the process of evaporation.

A participant asked whether students become more aware of their own acquisition of scientific processes through engagement with the curriculum models. Krajcik responded that the IQWST materials are extremely explicit about this. For example, when introducing the concept of scientific explanations, students are given a problem and a proposed explanation and invited to comment on the quality of the explanation. The materials explicitly describe what a scientific claim is, what constitutes evidence, how reasoning is used, and the role of each component in building a scientific explanation. Kolodner said that the LBD curriculum model uses the same approach, with explicit description of what a claim is and what constitutes evidence to support a claim. Students share their explanations with the class, discuss what makes one explanation better than another, and develop a whole-class explanation that they can all agree on. Krajcik said that the IQWST materials also engage students in publicly sharing their explanations and obtaining feedback from other students in order to improve their explanations. They encourage students to provide helpful feedback on other students' explanations, such as noting if an explanation lacks evidence or reasoning.

Reflecting on the session and the previous day's session on promising models (see Chapter 4), Parravano said that "there is very, very good indirect evidence . . . that these materials really are able to develop 21st century skills." He noted that all of the presenters had emphasized the importance of fidelity in delivery of the curriculum models, commenting that this finding underlined the importance of the upcoming workshop session on teacher readiness for 21st century skills, discussed in Chapter 6.

6

Science Teacher Readiness for Developing 21st Century Skills

This chapter addresses the workshop guiding questions focusing on science teachers: What is known about how prepared science teachers are to help students develop 21st century skills? What new models of teacher education may support effective teaching and student learning of 21st century skills, and what evidence (if any) is available about the effectiveness of these models? It summarizes a commissioned paper addressing these questions and the following discussion.

HOW TEACHER EDUCATION WILL HAVE TO EVOLVE

Mark Windschitl (University of Washington) presented a paper on science teacher readiness for cultivating 21st century skills (Windschitl, 2009). He opened with a comparison between the learning goals of reform in science teaching and the learning goals of 21st century skills, suggesting that most of the latter can be taught in the context of scientific inquiry or project-based learning. However, achieving this potential will require "ambitious" teaching, which:

- features learning how to solve problems in collaboration with others;
- engages students in productive metacognitive strategies about their own learning;
- places some learning decisions and activities in the hands of students that were formerly determined by the teacher; and

- depends for success on monitoring of student thinking about complex problems and relies on ongoing targeted feedback to students.

Windschitl warned that this type of ambitious teaching is unlike instruction in which most teachers have participated or even witnessed. Past efforts to reform teaching have had only a "modest track record," he said, and the broad trends in science classrooms today suggest that improvements are needed. Classes often focus on activity rather than sense-making discourse (Roth and Garnier, 2006, 2007; Weiss et al., 2003); teachers rarely press students for explanations, use questioning effectively, or take into account students' prior knowledge (Baldi et al., 2007; Banilower et al., 2008).

In the face of these disturbing trends, Windschitl said, it is important to consider what the research tells us about how teachers learn to teach science. First, content knowledge is very important, and is related to student learning (Magnusson et al., 1992). Teachers with strong content knowledge are more likely to teach in ways that help students construct knowledge, pose appropriate questions, suggest alternative explanations, and propose additional inquiries (Alonzo, 2002; Brickhouse, 1990; Gess-Newsome and Lederman, 1995; Lederman, 1999; Roehrig and Luft, 2004; Sanders, Borko, and Lockard, 1993). Second, he said, preservice teachers come into preparation with deeply engrained theories about what counts as good teaching and what counts as learning. These theories can be resistant to change and may filter out learning of new approaches to science instruction, unless teacher educators surface the theories and work actively to counter them.

Model Teacher Preparation, Induction, and Professional Development Programs

Teacher preparation programs capable of addressing these learning challenges have several characteristics, Windschitl said. They center on a common core curriculum grounded in substantial knowledge of child or adolescent development, learning, and subject-specific pedagogy. They provide students with extended opportunities to practice under the guidance of mentors (student teaching), lasting at least 30 weeks, that reflect the program's vision of good teaching and are interwoven with course work. Short-term interventions have shown little capacity to change teacher preconceptions (Wideen, Mayer-Smith, and Moon, 1998), but longer term approaches that explicitly seek to elicit and work with novice teachers' initial beliefs have shown some success in fostering reform-based teaching (Fosnot, 1996; Graber, 1996; Windschitl and Thompson, 2006). Other characteristics of effective teacher preparation programs include extensive use of case study methods, teacher research, performance assessments, and portfolio examinations that relate teachers' learning to classroom practice (Darling-Hammond, 1999).

In their first two years on the job, new teachers often are caught up in a frantic cycle of planning, teaching, and grading, with the result that they often shelve advanced teaching strategies developed in their teacher preparation programs. Windschitl said that induction programs can counter this cycle, providing an excellent opportunity to maintain a focus on 21st century skills in collaborative professional settings. One of the most promising practices for both induction and professional development involves bringing teachers together to analyze samples of student work, such as drawings, explanations, essays, or videotaped classroom dialogues. Based on principled analyses of how students are responding to instruction, the teachers change their instructional approaches. This collaborative analysis of evidence of student learning is used in several Asian nations whose students perform very well in international comparisons of mathematics and science achievement (Lewis and Tsuchida, 1997; Ma, 1999; Marton and Tsui, 2004; Yoshida, 1999).

Windschitl then identified several features of professional development that can support reform-based teaching and teacher understanding of how to cultivate 21st century skills:

- Active learning opportunities focusing on science content, scientific practice, and evidence of student learning (DeSimone et al., 2002);
- Coherence of professional development with teachers' existing knowledge, other development activities, existing curriculum, and standards in local contexts (DeSimone et al., 2002; Garet et al., 2001);
- The collective development of an evidence-based "inquiry stance" by participants toward their practice (Blumenfeld et al., 1991);
- The collective participation by teachers from the same school, grade, or subject area (DeSimone et al., 2002); and
- The importance of time needed for planning and enacting new practice.

Windschitl clarified that coherence with existing knowledge does not mean tailoring instruction to what teachers already know, but rather taking into account their deeply engrained theories about "good" teaching and learning. There is a broad consensus in the research, he said, that "reform-oriented" professional development (activities such as teacher study groups) results in more substantive changes in practice than "traditional" professional development (workshops or college courses) (Loucks-Horsley et al., 1998; Putnam and Borko, 2000). He then summarized his recommendations for teacher preparation, induction, and professional development (see Table 6-1).

Turning to his own research, Windschitl said the goal of the Teachers Learning Trajectory Initiative is to create systems capacity for continuous improvement in teachers' ability to foster 21st century skills. To learn more about how novices become experts, his research team followed 15 teachers for 3 to 4 years, through their preservice preparation and into their first or second year of teaching. In the preparation program, the future teachers were instructed in reform-based teaching, and, once on the job, they participated in an induction program focusing on review and analysis of student work. Over the course of the study, about one-third of the teachers developed "expert-like" teaching practice.

Windschitl reported that, when his team developed some "rudimentary tools" to assist the novice teachers, they were amazed at how well they improved their instruction (Windschitl, Thompson, and Braaten, 2009). The researchers hypothesized that the widespread use of the tools was attributable to the fact that they were tailored specifically to the needs of novices for planning, teaching, and assessment. For example, they observed that teachers were giving an assessment tool directly to their students to use in classroom conversations. It appeared that the teachers saw value in the tool and thought students could themselves benefit from it, by using the language in the tool to make their own judgment of their personal levels of explanation. This observation led the team to recognize that well-structured tools, especially those acting in a coherent system of support for ambitious teaching, could be very valuable. Based on this new understanding, the team

TABLE 6-1 Supports for the Teaching of 21st Century Skills

Element of Teacher Learning	Teacher Preparation	Induction	Professional Development
Characteristics	Deep, connected content knowledge	Not optional	Focuses on big pedagogical ideas
	Reframing of tacit, deeply engrained theories	Subject-matter specific	Includes time to plan for implementation
	Extended student teaching with master teacher, coherent with reform-oriented curriculum and 21st century skills	Focus on improving practice by examining evidence of student learning	Collective development of an inquiry stance to practice
		Builds on best practices from teacher preparation	Coherence with teachers' knowledge, school curricula

SOURCE: Windschitl (2009).

was funded by the National Science Foundation to develop such a system of tools.[1] Windschitl described the new suite of tools as follows:

1. Video-enhanced learning progressions for teachers, incorporating specific techniques of high-quality science instruction. For example, one tool illustrates three levels of increasing sophistication in the technique of pressing students for the evidence supporting their explanations.
2. "Big idea" tools, which help teachers take many different ideas presented in the curriculum and reconstruct them around a few big ideas. These tools could help foster nonroutine problem solving.
3. Rubrics to help teachers imagine certain kinds of student performance and to assess students' thinking, which was listed as a criterion in the rubric.
4. A suite of discourse tools to support teachers in developing complex communication skills. Windschitl described these tools as especially valuable in light of findings from the longitudinal study that teachers struggle with classroom discourse. One tool presents strategies to elicit students' initial hypotheses about important scientific ideas. Another focuses on ways to engage students in sense-making reflection on activities, and a third demonstrates how to press students for evidence-based explanations.
5. A set of tools and routines for teachers to use in collaboratively analyzing the effectiveness of their instruction, based on evidence of student learning.

Windschitl then described the challenges involved in moving toward teaching of 21st century skills (see Figure 6-1). He observed that the skills are not clearly defined, yet they call for "a fundamentally different vision of what counts as good teaching and what counts as learning." Developing expertise in teaching 21st century skills, he said, will require many years of coherent teaching, reflection, and professional development experiences that build on one another. He also said that efforts to promote such teaching will require reengineering of many interrelated components of the education system. Drawing an analogy between the education system and a food web made up of interdependent organisms, Windschitl asked whether there was any part of the education system that would not have to change, in order to foster students' 21st century skills. The answer, he replied, was no.

---

[1] See https://depts.washington.edu/mwdisc/.

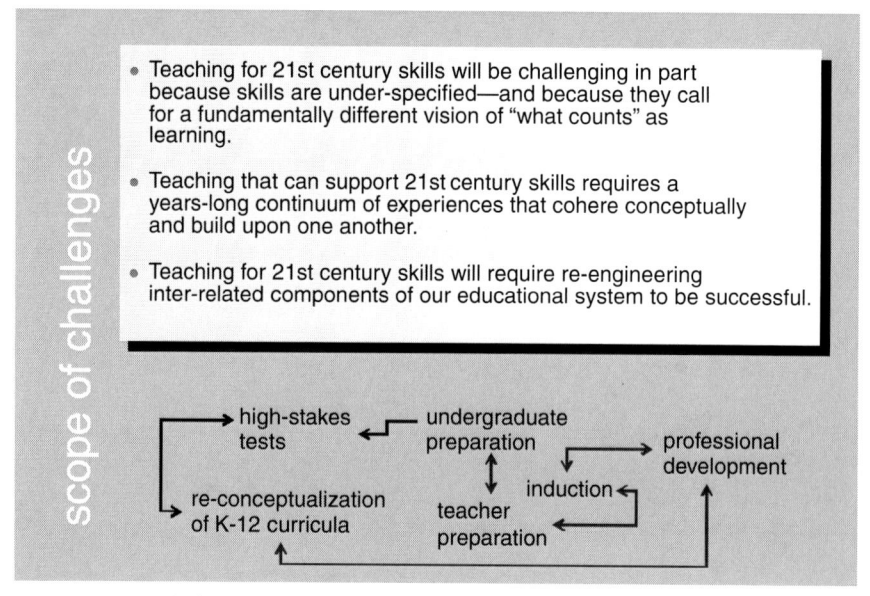

FIGURE 6-1 Challenges of teaching 21st century skills.
SOURCE: Windschitl (2009).

RESPONSE: THE VIEW FROM THE CLASSROOM

Elizabeth Carvellas (National Research Council) thanked Windschitl for his paper, saying its messages were very welcome after her many years of science teaching. Reflecting on the paper's summary of research knowledge about effective teacher professional development, preservice education, and ongoing support for teachers, she asked why this knowledge was not reaching teachers. In order to rapidly change teaching to develop 21st century skills, she said, it will be increasingly important for teachers to be able to easily access and apply research on teaching and learning. Next, she pointed out that teachers need time to prepare for these major changes in teaching. Some teachers, she noted, are responsible for teaching science to as many as 180 students. In the course of a school day, these teachers have three time periods for preparation, and deliver instruction to six classes. Although teachers are willing to teach in a different way, they need time and support to do so.

Carvellas then identified several other changes in the education system that she sees as necessary for many teachers to adopt 21st century teaching styles. First, she reminded audience members that they had heard the previous day about the importance of support from administrators. In addition to support, she suggested that administrators provide a guiding vision of 21st century teaching and learning. Second, she said, interdisciplinary work is required across the curriculum, not only in science. Third, she called for increased collaboration between science programs and teacher preparation programs in colleges and universities. She suggested that teachers could take the science content lessons from their science programs and use it in preservice education seminars and discussions with experts in child and adolescent development and learning, in order to translate the content "into something that works for kids."

Fourth, Carvellas said that science teachers, especially those who teach outside their field of undergraduate study, require ongoing support and professional development around the big ideas and concepts of science. In rural high schools, she said, a single teacher may be responsible for teaching chemistry, earth science, physics, and biology, requiring strong content knowledge of all four subjects. Fifth, after agreeing with Windschitl on the need for ongoing, long-term professional development, she proposed careful design of it to meet the needs of teachers in particular subjects with particular groups of students. She observed that many elementary and secondary school teachers are currently working hard to provide differentiated instruction to meet the needs of individual students (Tomlinson, 2003), yet these same teachers receive "one size fits all" professional development. Carvellas suggested that online teacher professional development might be the best way to support teachers in moving toward 21st century teaching, as discussed in a recent report by the National Academies' Teacher Advisory Council (National Research Council, 2007b). Finally, she expressed strong agreement with Windschitl about the value of engaging teachers in collaboratively analyzing the effectiveness of their instruction, based on evidence of student learning.

DISCUSSION

Reflecting on the presentation and response, moderator William Sandoval observed that both speakers called for a fundamental restructuring of teachers' daily schedules, with more time for planning and collaborative analysis of student work. Noting that such changes are currently taking place in only a handful of schools, led by a far-sighted principal or group of teachers, he asked the speakers how to make this kind of restructuring more systemic. Windschitl responded that this kind of major change requires new policies to convert teaching into a profession, rather than simply a job. Echoing earlier comments by Anderman (see Chapter 3), Windschitl said that policies in Asian countries recognize and support teachers as professionals. For example, he said, teachers in Japan and Singapore use lesson study to help plan, test, and revise lessons, and lesson study is "built into" their identity as teachers (Lewis and Tsuchida, 1997). Teachers in these nations have time off from instruction during the school day, so they can observe other teachers. In Singapore, Windschitl said, teachers can win a grant to support travel abroad to visit an outstanding school or teacher (Darling-Hammond and Cobb, 1995).

Carvellas suggested returning to the earlier workshop discussions that focused on thinking about education as a system (see Chapter 3). Because teachers are "part and parcel" of the system, she said, it is important to involve them and obtain their views about proposed changes. Sandoval replied that, as a researcher, he welcomes this advice, because it is always difficult to obtain the resources necessary to implement collaborative lesson study, and teachers can advise researchers on how to obtain these resources.

Following the panel discussion, Sandoval invited the workshop participants to use their notebooks to write down two concrete recommendations to support rapid development of 21st century teaching. After several minutes, he asked for volunteers to share their recommendations. One participant suggested starting high school classes an hour later, both to accommodate adolescent sleep schedules and to provide an hour of planning time to teachers. Another recommended changing undergraduate introductory science classes to include 21st century skills, as a model for future science teachers. Windschitl responded that changing undergraduate science courses would require a major reorganization of the curriculum, along with retraining of faculty members and other instructors. Carvellas observed that the large size of many undergraduate introductory science classes makes it difficult for instructors to engage students in discourse and develop their 21st century skills.

Bruce Fuchs offered a "radical" proposal to close half the schools of education, because, he argued, the annual number of new bachelor's and master's graduates with education degrees is greater than the number of vacancies. One result, in his view, is that people who never really wanted to become teachers end up in the classroom.

Jay Labov (National Research Council) suggested helping science graduate students, who will become the next generation of faculty, become aware of the research on undergraduate science learning and teaching. At the high school level, he said, the College Board is currently revising the Advanced Placement (AP) Program in response to a National Research Council report (2002), and these changes may support development of 21st century skills. Labov recommended engaging undergraduate science faculty, in collaboration with AP teachers, to consider how best to prepare AP teachers to deliver the innovative science curricula that develop 21st century skills (see Chapters 4 and 5).

Kenneth Kay offered two policy strategies that he said would complement the agenda for teacher preparation and professional development proposed by Windschitl. First, he proposed that every state adopt new teacher certification requirements incorporating 21st century skills, as

North Carolina has done. Second, he suggested that states and districts provide performance incentives to teachers who demonstrate the capacity to teach 21st century skills.

Eric Anderman agreed with Windschitl about the value of extended student teaching experiences, lasting at least 30 weeks, but called for improved monitoring of the teacher mentors who supervise the student teachers. He recommended that mentors be selected carefully and provided with monetary compensation, rather than continuing education credits. Carvellas heartily agreed with this suggestion, observing that expert teachers with 20 or more years of service do not need continuing education credits. She asked for improved compliance with existing guidelines that require that mentors do much more than simply "drop by once a week," adding that compensation for these mentors is critical.

Reflecting on the topic of mentoring student teachers, Sandoval mentioned the national problem of low teacher retention rates, as many teachers leave the profession after just a few years. Windschitl responded that current education policies often focus on producing new teachers, instead of retaining high-quality teachers. Many new graduates with education degrees, he said, are not prepared adequately in classroom management, in responding to linguistic and cultural diversity in the classroom, or in teaching science. As a result, he said, many leave teaching within 3 to 5 years.

Joyce Winterton (Office of Education, National Aeronautics and Space Administration) suggested that her agency collaborate with the National Institutes of Health, the U.S. Department of Energy, and the National Science Foundation to create externships for teachers. In these positions, teachers would participate in research projects at national laboratories and in industry.

Rodger Bybee questioned the usefulness of "radical" recommendations because, in his view, the education system will reject such sweeping change. He recommended instead building on the tools for teachers developed by Windschitl, which support smaller, more achievable change. Sandoval agreed that it is important to try to build on models of positive change.

Raymond Bartlett (Teaching Institute for Excellence in Science Technology, Engineering, and Mathematics) said, in his years of work in industry and with a state board of education, he learned that it is possible to make major changes in the education system. For example, a change in teacher certification requirements will dramatically change the whole system. He suggested that, rather than talking to each other about science education and 21st century skills, participants begin discussions with key organizations in Washington, DC, such as the Association of State Boards of Education, which are positioned to support and implement major changes in education policy.

7

Assessment of 21st Century Skills

This chapter summarizes two presentations and a discussion of the assessment of 21st century skills. The first section of the chapter focuses on methods used by some large corporations to assess the 21st century skills of current employees and job applicants. The second section summarizes a commissioned paper focusing on assessment of 21st century skills in educational settings. The final section of the chapter summarizes discussion of the presentation and paper.

CORPORATE ASSESSMENT

Janis Houston opened her presentation (Houston and Cochran, 2009) by highlighting the purposes of corporate assessment. She noted that the purposes for which any test is used (whether in education or in making employment decisions) affects the methods used in developing the test. In the world of employment, assessments are used for selection (to try to predict whether an individual will perform well in the future), for promotion, for certification (to ensure that an individual possesses a certain standard body of knowledge), and to identify training and development needs.

Houston then outlined the types of assessments used by large organizations, which include multiple-choice tests of cognitive abilities (e.g., mathematics) and noncognitive characteristics (e.g., personality type), structured interviews, situational judgment tests, role plays, group exercises, in-basket exercises, work samples, and performance standards/appraisal. She discussed three of these types in greater depth. All three are designed to assess a candidate's readiness for a specific job, which may include assessment of

21st century skills if such skills are important for successful performance in the job.

Role Play

In a role play assessment, the candidate for promotion is provided with written information about a realistic situation that may involve a nonroutine problem. After a period of time to prepare for the role play, the candidate presents her or his response to the situation to a panel of trained assessors. The assessors rate the response using behaviorally anchored rating scales, which describe specific behaviors.

For example, a candidate for promotion may be asked to play the role of a newly promoted insurance investigator who has just been put in charge of a large-scale insurance fraud investigation and is about to meet with a claims adjuster and an FBI agent. The candidate is told that it is important to ensure that the investigation is led by his or her company, rather than the FBI, and is given an hour to prepare for the 30-minute meeting. The managers conducting the assessment also prepare another person to participate in the role play, in the role of the FBI agent. This person is instructed to be very aggressive, to interrupt the candidate frequently, to push the candidate to turn the case completely over to the FBI, and to provide the candidate with new information about the criminal past of the suspects.

Houston explained that trained assessors use a scoring system to evaluate the candidate's adaptability, as displayed in the role play. This system guides assessors to award few points for adaptability if the candidate acts flustered or overwhelmed by new information and more points if the candidate seamlessly adjusts to new information.

Developing and administering role plays involves several challenges, Houston said. First, substantial input from subject-matter experts is necessary to identify appropriate problems or situations, and personnel testing experts are also needed to create the role play materials and behaviorally anchored ratings, so the process is labor-intensive. Second, because only one candidate can participate at a time, this form of assessment is expensive. Many role plays involve additional role players along with the candidate, and these other role players must be paid for their time. Finally, a role play can be scored only by engaging the time and expertise of multiple trained assessors. Unlike a typical multiple-choice test used in educational settings, a role play cannot be electronically scored.

Group Exercise

The development, administration, and scoring of a group exercise are similar to the role play, Houston said. The critical difference is that candidates work in groups to address a problem or respond to a situation, making it possible to assess their interactive skills, such as negotiation, persuasion, and teamwork.

Houston said that the costs and challenges involved in this form of assessment are similar to those in a role play, requiring the time of subject-matter experts and test developers. It must be carefully designed in order to manage the group interaction, and it often requires the time of other trained role players in addition to the candidates. Finally, scoring the group exercise requires the participation of multiple trained assessors; it cannot be electronically scored.

Houston noted that corporations are willing to pay the high costs ($250,000-$500,000) of creating role plays and group exercises, because these forms of assessment are well accepted by candidates, and they yield more information about a candidate's skills than a multiple-choice written test. These two forms of assessment, she said, help organizations to hire more highly qualified candidates, resulting in increased productivity and job performance.

Situated Judgment Tests

Houston said that a situated judgment test is less expensive than either a role play or a group exercise. In this test, the candidate is presented with a realistic hypothetical situation and a list of five to eight possible responses to the situation. The candidate may be asked to select the most and least effective option or only the most effective option. The scoring of the test is based on the effectiveness of the options the candidate selects.

Houston presented an example of a situated judgment test designed to assess adaptability. The candidate is presented with a situation in which he or she is working intensely to finish a report for his or her supervisor, when the supervisor calls to say he or she would like to touch base about the candidate's progress on another project in one hour. The optional responses range from getting as much of the report done as possible within the hour to explaining to the supervisor about the unfinished report and asking for an extension on the report deadline.

Houston said this type of test is much less expensive to administer than either a role play or a group exercise. The test is developed by identifying the skills to be assessed, creating realistic situations or problems in consultation with subject-matter experts, generating multiple response options for each situation, and devising a scoring system to determine the effectiveness of each optional response. The test situations can be presented either in print or by video, and the candidate responds in writing. Like educational assessments, the situated judgment test can be administered to a large group simultaneously and can be electronically scored.

Conclusion

Houston concluded by saying that development and administration of all three types of corporate assessments is a labor-intensive, expensive process that may not be practical for use in large-scale educational assessment. However, relative to the role play and the group exercise, situated judgment tests are far easier to administer and score; they can easily be administered to large groups and scored by computer.

TOWARD A FRAMEWORK FOR ASSESSING 21ST CENTURY SCIENCE SKILLS

Maria Ruiz-Primo (University of Colorado Laboratory for Educational Assessment, Research, and Innovation) presented a summary of her paper on assessment of 21st century skills (Ruiz-Primo, 2009).

Model of Assessment Development

Ruiz-Primo observed that her approach to developing a framework to assess 21st century skills in the context of science is based on a theoretical model for the development and evaluation of assessments, which takes the form of a square. The first step in the model is to define the construct—the knowledge, skills, or other attributes to be assessed. Based on this definition, the developers use conceptual analysis to identify behaviors, responses, or activities most representative of the construct in order to create an observation model. Next, the developers use the observation model as the basis for developing the assessment, with specific situations designed to elicit the behaviors, responses, or activities included in the observation model. After administering the assessment, the developers conduct empirical analysis to interpret the results and analyze whether the evidence collected supports the inferences about the knowledge, skills, or other attributes of the construct. The result of this analysis may lead to revision of the construct, the observation model, or the assessment itself.

Defining the Construct

Ruiz-Primo began by defining the construct of 21st century skills in the context of science. She identified dimensions of the five skills and the research underlying each dimension. She then compared these dimensions with three other recent models: (1) the framework developed by the Partnership for 21st Century Skills (2009b); (2) the Standards for the 21st Century Learner of the American Association of School Librarians (2009); and (3) the enGauge 21st Century Skills developed by the North Central

Regional Education Laboratory and the Metiri Group (Lemke et al., 2003). She observed that the other three models most strongly emphasized communication and nonroutine problem-solving skills.

Turning to the context of science education, Ruiz-Primo compared dimensions of the five skills with the definition of science proficiency developed in a recent Board on Science Education review of the research on science learning in grades K-8 (National Research Council, 2007a). The review concluded that students who are proficient in science should:

1. know, use, and interpret scientific explanations of the natural world;
2. generate and evaluate scientific evidence and explanations;
3. understand the nature and development of scientific knowledge; and
4. participate productively in scientific practices and discourse.

Ruiz-Primo found that dimensions of two of the five 21st century skills—complex communication/social skills and nonroutine problem solving—were most closely aligned with this definition of science proficiency.

Proposed Construct

Building on these two analyses, Ruiz-Primo proposed a construct of 21st century skills in the context of science that includes three domains:

1. Dispositions, general inclinations or attitudes of mind;
2. Cross-functional skills (cognitive skills that are likely to be used in any domain); and
3. Science knowledge.

She included two of the 21st century skills—adaptability and self-management/self-development—in the domain of dispositions, and two others—complex communication/social skills and nonroutine problem-solving skills—in the domain of cross-functional skills. The science knowledge domain is defined by the four strands of science proficiency listed above (National Research Council, 2007a). The resulting construct is represented in Figure 7-1.

Types of Science Knowledge

Expanding her analysis of science content, Ruiz-Primo explained that her research group has proposed an approach to understanding and developing measures of science achievement based on the idea of types of

Dispositions	Cross-Functional Skills	Content (Scientific Proficiencies)
Adaptability & Self-Management	Problem Solving & Complex Communication	-Know, use, and interpret scientific explanations of the natural world -Generate and evaluate evidence and explanations -Understand the nature and development of scientific knowledge -Participate productively in scientific practices and discourse

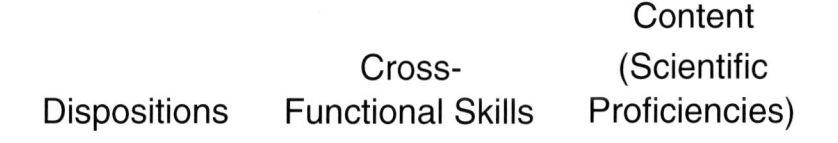

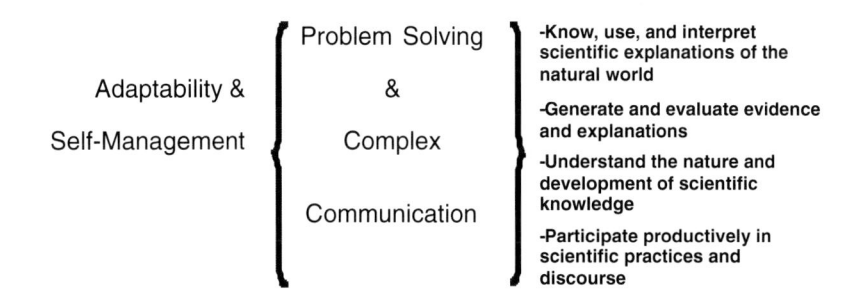

FIGURE 7-1 Construct domains of 21st century skills in the context of science education.
SOURCE: Ruiz-Primo (2009).

knowledge (Li, 2001; Li, Ruiz-Primo, and Shavelson, 2006; Ruiz-Primo, 1997, 1998, 2003; Shavelson and Ruiz-Primo, 1999). They include

- Declarative knowledge: knowing that. This type includes knowledge that ranges from discrete and isolated content elements, such as terminology, facts, or specific details, to a more organized knowledge forms, such as statements, definitions, knowledge of classifications, and categories.
- Procedural knowledge: knowing how. This type involves knowledge of skills, algorithms, techniques, and methods. It usually takes the form of if-then production rules or a sequence of steps (e.g., measuring temperature using a thermometer; applying an algorithm to balance chemical equations; adding, subtracting, multiplying, and dividing whole numbers).
- Schematic knowledge: knowing why. This type involves more organized bodies of knowledge, such as schemas, mental models, or "theories" (implicit or explicit) that are used to organize information in an interconnected and systematic manner.
- Strategic knowledge: knowing when, where, and how to apply knowledge. "The application of strategic knowledge involves navi-

gating the problem, planning, monitoring, trouble-shooting, and synchronizing other types of knowledge. Typically, strategic knowledge is used when one encounters ill defined tasks" (Li and Tsai, 2007, p. 14).

Ruiz-Primo proposed that this typology of knowledge has three important implications for science assessment. First, it can be applied to determine what types of knowledge are being measured by a particular assessment; second, it can be used to interpret student scores; and third, it can be applied to design or select assessment tasks that are aligned with instructional goals.

An Approach to Developing and Evaluating Assessments

Ruiz-Primo explained that defining the construct represents the first step in developing and evaluating an assessment. The next steps include development of observation models, specifying those aspects of a student's response to a test item that would be valued as evidence of the construct, and linking the observation models with the design of the assessment.

Retrospective Logical Analysis

Assessment researchers use retrospective logical analysis to analyze assessment tasks that have already been developed. In this type of analysis, they review how the task elicits the targeted knowledge and influences students' thinking and responses. Ruiz-Primo identified four criteria that can be applied in retrospective logical analysis:

1. Task demands: what students are asked to perform (e.g., define a concept or provide an explanation);
2. Cognitive demands: inferred cognitive processes that students may act on to provide responses (e.g., recall a fact or reason with a model);
3. Item openness: the extent of constraints in the response (e.g., selecting versus generating responses or requiring information only found in a task versus information that can be learned from the task); and
4. Complexity of the item: the diverse characteristics of an item, such as familiarity to students, reading difficulty, and the extent to which it reflects experiences that are common to all students.

Returning to the goal of assessing complex communication and nonroutine problem solving, Ruiz-Primo said that the assessment items should be designed to yield evidence of the solving of complex problems. She then identified several dimensions related to the complexity of a problem (see Table 7-1). One is the structure of a problem, which is determined by the test developer. A problem may be well structured or ill structured. Another dimension is whether the problem is routine or nonroutine, and this depends on whether the examinee has already learned procedures to solve this type of problem. In another dimension, a "rich" problem requires activities and subtasks, while a lean problem does not. Yet another dimension is the extent to which it requires prior exposure to the topic in the context of school. Other dimensions include whether the solution is approached individually or in collaboration with others and whether there is a time constraint to solve the problem. Finally, she observed that a problem may vary in the extent of communication required to respond, with a selected or short-answer item requiring less writing and a constructed-response item

TABLE 7-1 Dimensions of Problem Complexity

Dispositions	Cross-Functional Skills			General Description
	1. Nonroutine Problem Solving			
	Nature of a Problem			
	Well structured	⟷	Ill structured	**Structure.** Level of problem definition. Defined by the developer
	Routine	⟷	Nonroutine	**Routine.** Solution procedures already learned? Depends on the examinee
Adaptability and Self-Management	Lean	⟷	Rich	**Richness.** Number of activities and assessment subtasks involved
	Schooled	⟷	Unschooled	**Schoolness.** Obligatory academic exposure to solve the problem?
	Independent	⟷ OR	Collaborative	**Collaboration.** Solution approached individually or with others?
	Untimed	⟷	Timed	**Time.** Is the time to solve the problem constrained?
	2. Complex Communication			
	Extent of Communication			
	Selected or short-answer	⟷	Constructed-response	Intensity or amount of writing

SOURCE: Ruiz-Primo (2009).

requiring more writing. Ruiz-Primo integrated all of these dimensions into a framework for retrospective analysis of existing assessment tasks, which she applied to review sample assessment tasks.

Review of Sample Assessment Tasks

Ruiz-Primo explained that she reviewed several existing assessment items to consider their ability to measure adaptability, complex communications, nonroutine problem solving, and self-management/self-development. She selected assessment items from science as well as other domains, as requested by the planning committee. Returning to the construct she proposes, she asked what types of assessment tasks would yield evidence that a student can "generate and evaluate scientific evidence and explanations" (National Research Council, 2007a, p. 2). In answer to her own question, she said that tasks should require students to do something or critique examples of scientific evidence and explanations, and they should involve ill-structured or nonroutine problems. In addition, she proposed that the tasks should be constrained in terms of time allowed or collaborations required, in order to measure not only complex communication and nonroutine problem solving, but also adaptability and self-management. Based on all of these considerations, she concluded that such assessment methods as essays and performance assessment tasks were promising candidates to assess the four 21st century skills. She then reviewed four assessment items, discussed below.

Analytic Writing Task

Ruiz-Primo described a writing task from the Collegiate Learning Assessment (CLA). Developed by the Council for Aid to Education and the RAND Corporation, the goal of the CLA is to measure "value added" by educational programs in colleges and universities. The writing task Ruiz-Primo examined presents a real-life scenario and allows students 30 minutes to construct a written argument for or against the principal's decision to oppose the opening of fast-food restaurants near the school (see Box 7-1).

Ruiz-Primo concluded that this problem is not well structured, increasing its complexity, because students are not able to determine what an acceptable answer would be or how to arrive at that answer. She said it is unclear whether the problem is routine or nonroutine; students being examined may or may not have already learned a routine procedure to approach it. The problem may require subtasks, such as making a list of pros and cons first before writing the argument. It does not appear to require an academic procedure taught at the school. Although CLA administers this type of task individually, it could also be a collaborative task. Because it

> **BOX 7-1**
> **Sample CLA Analytic Writing Task: Critique an Argument**
>
> A well-respected professional journal with a readership that includes elementary school principals recently published the results of a two-year study on childhood obesity. (Obese individuals are usually considered to be those who are 20 percent above their recommended weight for height and age.) This study sampled 50 school children, ages 5-11, from Smith Elementary School. A fast-food restaurant opened near the school just before the study began. After two years, students who remained in the sample group were more likely to be overweight relative to the national average. Based on this study, the principal of Jones Elementary School decided to confront her school's obesity problem by opposing any fast-food restaurant openings near her school.
>
> SOURCE: Collegiate Learning Assessment Fall 2008 Interim Report. Reprinted with permission from the Council for Aid to Education.

is timed, students need to self-manage their time to finish it in 30 minutes. The context of the task can be considered "social," something that students might observe in their own community.

According to the CLA framework, the test measures skills that are applicable to a wide range of academic subjects and are also valued by employers, including critical thinking, analytic reasoning, problem solving, and written communication (Klein et al., in press). This particular item emphasizes written communication. Ruiz-Primo noted that the rubric used to score the item indicates that student performance is evaluated based on dimensions of the four strands of science proficiency (National Research Council, 2007a), including evaluation of evidence, analysis and synthesis of evidence, and drawing conclusions as well as on criteria related to the quality of the writing.

Based on this analysis, Ruiz-Primo recommended considering similar tasks for assessing 21st century skills.

Performance Task

The next example, also from CLA, was a 90-minute performance task. This task invites students to pretend that they work for a company and that their boss has asked them to evaluate the pros and cons of purchasing an airplane (called the SwiftAir 235) for the company. The task indicates that concern about this purchase has risen with the report of a recent SwiftAir 235 crash. Students are invited to respond in a real-life manner by writing a memorandum (the response format) to their boss analyzing the pros and cons of alternative solutions, anticipating possible problems and solutions to them, recommending what the company should do, and focusing on evidence to support their opinions and recommendations. The scoring of student performance on this item recognizes and evaluates alternative justifiable solutions to the problem and alternative solution paths.

Ruiz-Primo said the problem is ill defined, lacking clear information on the characteristics of the correct response. Although routine procedures to solve the problem depend on the knowledge that the examinee brings to the situation, many examinees will have some sense of how to approach it, such as to read the information provided. It is a rich problem, since examinees are provided detailed information about the SwiftAir 235, in particular, and airplane accidents, in general. Because some of the provided information is relevant and sound, while some is not, part of the problem involves determining what is relevant. The problem does not appear to require academic exposure to solve it, and it is an individual problem. Because the problem is timed, students must apply self-management skills. Finally, the problem context is job related.

Science Achievement Task

Next, Ruiz-Primo presented a task from the Program for International Student Assessment 2006 science test (see Box 7-2), referred to as the School Milk Study. Both of the questions are designed to elicit some of the behaviors, responses, and actions defined in the observation (evidence) models for the 21st century science skill to "generate and evaluate evidence and explanations," (National Research Council, 2007a, p. 2).

She explained that the item is well structured, with one correct response, reducing its complexity. As with the previous examples, the extent to which the problem is routine or nonroutine may vary, depending on how much experience a given student has in identifying scientific problems and using evidence to support explanations. The problem does not require students to carry out different activities or carry out subtasks and is based on processes and procedures learned in school. It is designed for individual response, and it is timed in relation to other items in the larger test. The item has a historical setting in a global context, and it demands no written communication.

She concluded that the item is designed primarily to assess declarative knowledge in the domain of science. Because the item is constrained, requiring selection of a correct response, it does not require the student to display complex written communication skills. The constraints reinforce the task and cognitive demands placed on students.

> **BOX 7-2**
> **The School Milk Study**
>
> In 1930, a large-scale study was carried out in the schools in a region of Scotland. For four months, some students received free milk and some did not. The head teachers in each school chose which of their students received milk. Here is what happened:
>
> - 5,000 school children received an amount of unpasteurized milk each school day.
> - Another 5,000 school children received the same amount of pasteurized milk.
> - 10,000 school children did not receive any milk at all.
>
> All 20,000 children were weighed and had their heights measured at the beginning and the end of the study.
>
> **Question 1: SCHOOL MILK STUDY**
>
> Is it likely that the following questions were research questions for the study? *Circle "Yes" or "No" for each question*
>
> | Is it likely that this was a research question for the study? | Yes or No? |
> | --- | --- |
> | What has to be done to pasteurize milk? | Yes/No |
> | What effect does the drinking of additional milk have on school children? | Yes/No |
> | What effect does milk pasteurization have on school children's growth? | Yes/No |
> | What effect does living in different regions of Scotland have on school children's health? | Yes/No |
>
> **Question 2: SCHOOL MILK STUDY**
>
> On average, the children who received milk during the study gained more in height and weight than children who did not receive milk.
>
> One possible conclusion from the study, therefore, is that school children who drink a lot of milk grow faster than those who do not drink a lot of milk.
>
> To have confidence in this conclusion, indicate one assumption that needs to be made about these two groups of students in the study.
>
> ---
>
> SOURCE: Organisation for Economic Co-operation and Development, Program for International Student Assessment (2006, p. 34). Reprinted with permission.

A Technology-Rich Science Task

Ruiz-Primo then presented a sample item that was field-tested in the National Assessment of Educational Progress (NAEP) Technology-Based Assessment Project (Bennett et al., 2007). Focusing on physical science, the item includes three problems, asking the student to determine how different payload masses affect the altitude of a balloon. Students are presented with a search scenario requiring them to locate and synthesize information about a scientific helium balloon, and a simulation scenario requiring them to experiment to solve problems. Students can see animated displays after manipulating the mass carried by the balloon and the amount of helium contained in the balloon.

Ruiz-Primo said that the problems are well structured, with a clearly correct answer. The problem appears routine, as most students know how to gather information from the World Wide Web. The simulation scenario is a fixed procedure that seems to require changing the values of the different variables; what is important is which values to select on each trial. The problems are rich, involving several activities and subtasks. Although it involves technical terms, the problem may not require previous learning of a specific procedure. The problem is from a science context, rather than a real-world context, and is to be solved individually within time constraints.

She concluded that this item taps mainly procedural knowledge, as students are asked to carry out procedures to search for information on the World Wide Web or to conduct the simulation. In addition, students do not have a choice about how best to represent the data collected; it is given. The item is constrained, requiring students to select from a set of tools, and these constraints reinforce the task and cognitive demands placed on students. Finally, it requires some written communication.

Measuring Strategic Knowledge

Noting that a review of the TIMSS 1999 Science Booklet 8 test identified no items that measure strategic knowledge (Li, Ruiz-Primo, and Shavelson, 2006), Ruiz-Primo said that computer technology now makes it possible to track the strategies students use when applying information to solve problems. She pointed to the example of the Interactive Multi-Media Exercises (IMMEX; Case, Stevens, and Cooper, 2007; Cooper, Stevens, and Holme, 2006), a system that presents students with real-world complex science problems to solve in an online environment. The program has been used to assess science learning in K-12, undergraduate, and medical education. It tracks students' actions and data-mining strategies to arrive at the solution, grouping the strategies into types and identifying pathways into specific strategy types. It provides reliable and repeatable measures of students' problem-solving skills (Case, Stevens, and Cooper, 2007; Cooper, Sandi-Urena, and Stevens, 2008). In addition, because it offers the possibility for collaborating to solve a problem, the system may be used to elicit students' communication skills, skills in considering others' opinions, adaptability, and self-management.

Conclusion

On the basis of her analysis of these items, Ruiz-Primo concluded that similar tasks should be considered for assessing 21st century skills. She offered four recommendations for future research and development of assessments of 21st century skills. First, she said, it is important to more carefully define the 21st century skills of interest. Second, for the purposes of developing large-scale assessments, it is important to identify the most critical skills, as she did when focusing on nonroutine problem solving and complex communication. Third, she said it is important to define the purposes of assessments designed to measure 21st century skills, such as to provide information for school accountability, to evaluate individual student progress, to focus public attention on educational concerns, or to change educational practices by influencing curriculum and instruction. She observed that different purposes require different sources of evidence to evaluate the validity of the assessment. Fourth, she said that computer-based technology can support the development, administration, and scoring of large-scale assessments of 21st century skills.

DISCUSSION

Session moderator Marcia Linn thanked Ruiz-Primo and Houston, observing that they had posed important questions about how to define the construct of 21st century skills, as well as how to measure this construct. She observed that Houston had demonstrated the importance of the goal of assessing 21st century skills by showing how much money private firms are willing to invest in assessments of these skills, as well as the cost savings that result from the use of these assessments.

She suggested research to develop work samples representing students' ability to apply scientific knowledge to every aspect of their lives. Noting that the presenters had suggested using technology to assess 21st century skills, Linn said that technology offers opportunities for synergy between the curriculum and the assessment. For example, she said, in her research team's online instruction, many of the student activities could be used as indicators of their 21st century skills. Linn suggested that teachers could score new types of science assessments capable of measuring 21st century skills. She said that, in the Netherlands, where complex assessments are used, schools send the completed assessments to other schools for grading. This process offers learning opportunities for teachers as well as students.

Linn then invited the audience to write down their reflections and questions about the session in their carbonless notebooks. After several minutes, she invited the audience to pose questions about the session and also to recommend policies or programs to support development of 21st century skills in science education.

The speaker and other workshop participants offered the following responses:

- Some states have already adopted educational standards incorporating 21st century skills and are beginning to develop assessments aligned with these new standards.
- Educational assessments are standardized for all students, in contrast to the process in the corporate world, which involves tailoring each assessment to a particular workplace or job.
- Corporate assessment must be tailored in order to be most relevant for the job and to be the most valid predictor of future job performance.
- In education, if the purpose is large-scale assessment, the same standardized test should be administered to all students. However, if a teacher wants to know how well her students have learned following a unit of instruction focused on 21st century skills, it may be appropriate to create a unique assessment.
- Although educational assessments measure individual skills, the value of a skill such as adaptability may be realized in groups, rather than as an attribute of separate individuals. From this perspective, it is important to think about assessment of people in groups.
- Online assessments can be manipulated to engage students in solving a problem with others who are not physically present. This approach could be used to assess a dimension of self-management—the ability to work as a member of a virtual team (Houston, 2007). Assessments can also be manipulated to change the status of the test-taker in order to assess adaptability. For example, a student may initially be asked to solve the problem individually, and then be told to collaborate with other students. This change of status could be used to assess adaptability, collaboration, and complex communication skills.
- In corporate assessment, the goal is for each individual to possess adaptability and other 21st century skills, as well as for groups to have these skills. Job performance tests of adaptability are sometimes used to identify those individuals who may be best able to cope with, and adapt to, physically dangerous situations.

8

Synthesis and Reflections

This chapter opens with a synthesis of the evidence of areas of intersection between science education and the development of 21st century skills. It then summarizes participants' reflections on the workshop and final comments from the planning committee members.

SYNTHESIS AND COMMON THEMES

In the following discussion, which is organized around the five 21st century skills, the preliminary definition of each skill is followed by a description of how the different presenters interpreted the relationship between the skill and the curriculum models they reviewed. The section concludes by identifying common themes that emerged across the curriculum models as well as in other workshop presentations.

Adaptability

Adaptability is defined as the ability and willingness to cope with uncertain, new, and rapidly changing conditions on the job, including responding effectively to emergencies or crisis situations and learning new tasks, technologies, and procedures. Adaptability also includes handling work stress; adapting to different personalities, communication styles, and cultures; and physical adaptability to various indoor or outdoor work environments (Houston, 2007; Pulakos et al., 2000).

The 5E Model

Bybee noted that although the integrated sequence of instructional activities in the 5E model may support adaptability, he did not find evidence of development of this skill in the research on the 5E model.

Online Learning Environments for Argumentation

Clark reviewed four online learning environments, each of which engages students in developing, warranting, and communicating a persuasive argument and in critiquing arguments developed by others. He proposed that these learning activities may support development of adaptability in three ways. First, the environments may help students adapt their everyday communication skills to align more closely with the skills used in scientific argumentation. Research on implementation of all four environments yields evidence that students improved either in scientific argumentation (Clark, 2004; Clark and Sampson, 2005; Clark et al., 2008; Cuthbert, Clark, and Linn, 2002) or in similar forms of argumentation (Janssen, Erkens, and Kansellar, 2007; Janssen et al., 2007; Marttunen and Laurinen, 2006; Salminen, Marttunen, and Laurinen, 2007; Stegmann et al., 2007; Stegmann, Weinberger, and Fischer, 2007).

Second, Clark proposed that adaptability develops as an offshoot of gains in argumentation, as students learn how to adapt to changing information or changing contexts. Research on some of the environments provides evidence of such development. For example, studies of the Dialogical Reasoning Educational Web Tool (DREW) indicate that it helps students learn how to identify and evaluate the arguments for and against a particular position when investigating an unfamiliar topic (Marttunen and Laurinen, 2006; Salminen, Marttunen, and Laurinen, 2007). Research on the Computer-supported Argumentation Supported by Scripts-experimental Implementation System (CASSIS) indicates that it is effective in improving students' ability to generate persuasive and convincing arguments and counterarguments (Stegmann et al., 2007; Stegmann, Weinberger, and Fischer, 2007).

Third, he suggested that these online environments develop adaptability by distributing and redistributing roles and activities to individual group members, so that they must take on new perspectives that may differ from their personal views and respond accordingly. For example, CASSIS uses scripts that guide learners to take on and rotate the roles of case analyst and constructive critic. Research indicates that the use of the scripts increased students' ability to elaborate arguments and counterarguments and share their knowledge and perspectives in the discussions (Weinberger, 2008; Weinberger et al., 2005).

These three strands of evidence suggest that the online argumentation environments develop students' adaptability to uncertain, new, and rapidly changing conditions.

Learning by Design

Kolodner noted that, in Learning by Design (LBD), students work on a variety of different teams throughout the school year, requiring them to adapt to fellow students with different working styles, strengths, and weaknesses. In addition, their work on multiple large challenges, each requiring different knowledge and skills, may develop adaptability.

Kolodner identified some evidence of development of adaptability in her comparison studies of LBD students and matched comparison classrooms, with students with similar levels of science achievement and socioeconomic status (Kolodner, Gray, and Fasse, 2003). The authors designed a performance task to assess the ability of student groups to design an experiment and gather and analyze experimental data. They also used written pre- and posttests, consisting mostly of multiple-choice items, to assess the students' content learning. They found that the LBD students immediately got down to work on the performance task, while the students in the comparison classroom took more time to get into groups and begin addressing the task. In comparison to the non-LBD students, the LBD students displayed significantly higher levels of negotiations during collaboration and distribution of the task among group members (Kolodner et al., 2003; Kolodner, Gray, and Fasse, 2003). These findings suggest that LBD students developed adaptability to new performance tasks and different personalities, communication styles, and cultures.

Investigating and Questioning our World through Science and Technology

Krajcik and Sutherland (2009) observed that the Investigating and Questioning our World through Science and Technology (IQWST) curriculum challenges students to build and revise models throughout the middle grades, based on evidence related to scientific phenomena. He argued that students' realization that models can change when new evidence is presented represents development of adaptability. In addition, he suggested that adaptability is supported by another IQWST learning goal—the expectation that, as learners become more sophisticated in constructing explanations, they will rule out other possible explanations. Krajcik observed that students need to consider if they have sufficient and appropriate evidence to support their claims and, if they do not, adapt by writing new claims that are supported by their evidence.

Research on implementation of IQWST provides evidence that students improve in modifying and revising models when presented with new evidence (Merritt, Shwartz, and Krajcik, 2008; Schwarz et al., 2009; Shwartz et al., 2008). The research also indicates that students improve in using evidence and reasoning to support their claims (McNeill and Krajcik, 2008a, 2008b; McNeill et al., 2006).

Complex Communication/Social Skills

Complex communication skills are defined as the ability to process and interpret both verbal and nonverbal information from others in order to respond appropriately. A skilled communicator is able to select key pieces of a complex idea to express in words, sounds, and images, in order to build shared understanding (Levy and Murnane, 2004). Skilled communicators negotiate positive outcomes with customers, subordinates, and superiors through social perceptiveness, persuasion, negotiation, instructing, and service orientation (Peterson et al., 1999).

The 5E Model

Bybee proposed that science curricula based on the 5E model support development of communication skills by engaging students in scientific argumentation.

He found evidence for development of argumentation in a study comparing an inquiry science curriculum based on the 5E instructional model with "commonplace teaching" of the same material, as defined by national surveys of science teachers (Wilson et al., 2009). The authors used a randomized controlled trial research design, assigning a total of 58 students ages 14-16 to either a group that was instructed based on the 5E model or to a group that received commonplace teaching. Students in the 5E group reached significantly higher levels of achievement compared with the other group in terms of three different learning goals—knowledge, scientific reasoning, and argumentation. The finding held for testing immediately following instruction and four weeks later.

Bybee also indicated that evidence of development of complex communication skills was provided by a study that found that the 5E model increased levels of higher order thinking among a small group of science students (Boddy, Watson, and Aubusson, 2003).

Online Learning Environments for Argumentation

Clark proposed that the online argumentation environments may support complex communication skills in at least two ways. First, all of the environments require students to develop, warrant, and communicate a persuasive argument, based on evidence. Because the goal of argumentation is to persuade and build shared understanding, "argumentation skills are therefore an integral component of complex communication skills" (Clark et al., 2009, p. 18). From this perspective, studies showing that students engaged with all four of these environments advance in argumentation may be seen as evidence of development of complex communication skills (e.g., Clark, D'Angelo, and Menekse, in press; Marttunen and Laurinen, 2006; Stegmann, Weinberger, and Fischer, 2007).

One way in which these environments support complex communication skills, Clark said, is through the use of scripts that orchestrate and control students' interactions with each other and the learning environment. He cited a study demonstrating that the use of scripts in CASSIS improved the quality of argumentation (Stegmann, Weinberger, and Fischer, 2007).

Learning by Design

Kolodner observed that the LBD curriculum integrates learning activities with opportunities for guided reflection that involve communication. Students present design ideas to each other, along with the scientific principles and experimental results that support these ideas. Teachers lead whole-class discussions after these sessions to help students focus on the ways in which science concepts are applied in the designs. Students also present their experimental procedures and results to each other in poster sessions, and the teacher leads several types of whole-class discussions. All of these activities have the potential to develop complex communication skills. In addition, an explicit goal of the curriculum is to enact a set of values and expectations, one of which is collaboration (Kolodner, Gray, and Fasse, 2003).

The comparison study described above found that the LBD students displayed significantly higher levels of negotiations during collaboration and distribution of the task among group members (Kolodner et al., 2003; Kolodner, Gray, and Fasse, 2003). Negotiation is a dimension of complex communication skills.

Investigating and Questioning our World through Science and Technology

Krajcik noted that several learning activities in IQWST may support development of complex communication skills. Students use evidence and reasoning to support claims (scientific explanations), both verbally and in writing; they also evaluate and critique claims made by others, both verbally and in writing. In addition, students construct models, present their models to class members, and justify their models based on evidence. In constructing models and writing scientific explanations, they need to consider that all the evidence is accounted for. Often, students construct explanations and build and revise models in small groups, and the groups present their explanations and models to other students for critique and feedback. These small group discussions, too, may support development of complex communication skills (Krajcik and Sutherland, 2009).

Several recent studies of implementation of IQWST indicate that students improve in these activities. These studies provide evidence that engagement with IQWST improves students' ability to support claims using evidence (McNeill and Krajcik, 2008a, 2008b; McNeill et al., 2006) and their ability to construct and communicate scientific explanations (Krajcik et al., 2008; Krajcik, McNeill, and Reiser, 2008). In addition, studies show that IQWST students improve in accounting for all evidence when constructing models (Merritt, Shwartz, and Krajcik, 2008; Shwartz et al., 2008). These findings may represent development of students' ability to select key pieces of a complex idea to express in words, sounds, and images, in order to build shared understanding.

Nonroutine Problem Solving

The ability to solve nonroutine problems is defined as follows: A skilled problem solver uses expert thinking to examine a broad span of information, recognize patterns, and narrow the information to reach a diagnosis of the problem. Moving beyond diagnosis to a solution requires knowledge of how the information is linked conceptually and involves metacognition—the ability to reflect on whether a problem-solving strategy is working and to switch to another strategy if the current strategy isn't working (Levy and Murnane, 2004). It includes creativity to generate new and innovative solutions, integrating seemingly unrelated information, and entertaining possibilities others may miss (Houston, 2007).

The 5E Model

Bybee identified three sources of evidence that engagement with the 5E model supports development of nonroutine problem solving. First, the comparative study discussed above (Wilson et al., 2009) measured student progress toward three goals of inquiry-based instruction, one of which was scientific reasoning. The study provides evidence that engagement with the 5E model increases students' scientific reasoning ability in comparison to the reasoning ability of students receiving more typical forms of science instruction. Bybee noted "a linkage between scientific reasoning and problem solving" (Bybee, 2009, p. 15). Second, he mentioned a study that found increases in students' higher order thinking following instruction based on the 5E model (Boddy, Watson, and Aubusson, 2003). Finally, he mentioned a study by Taylor, Van Scotter, and Coulson (2007), which found that students whose teachers fully implemented the 5E model were more able to apply their understanding to new situations than students whose teachers did not fully implement the model.

Online Learning Environments for Argumentation

Clark discussed the role of online argumentation environments in development of nonroutine problem solving. In WISE, students negotiate consensus and critique novel ideas rapidly introduced by other students, requiring them to examine and use a broad span of information from the initial laboratory activities, simulations, and everyday experiences. Research on students using this environment suggests that these activities increase the conceptual and structural quality of students' argumentation (Clark, D'Angelo, and Menekse, in press). This research suggests that students improve in such elements of nonroutine problem solving as integrating seemingly unrelated information and entertaining possibilities others may miss.

The CASSIS environment encourages learners to apply attribution theory to solve authentic problems, with support from argumentative collaboration scripts. In recent studies, CASSIS learners supported with an epistemic script were better able than other CASSIS learners to focus on the core aspects of a problem case and also pursued additional information and explored multiple perspectives (Mäkitalo et al., 2005; Weinberger, 2008; Weinberger et al., 2007). These findings suggest that engagement with CASSIS supports development of such dimensions of nonroutine problem solving as learning to analyze large amounts of information, recognize patterns, and determine whether or not a claim is well supported by available evidence.

The argument diagram tool in the DREW environment can also promote nonroutine problem-solving skills. The results on DREW thus far have shown that students deepen and broaden their knowledge of a given topic when diagrams are used across three sequential phases of students' work (Marttunen and Laurinen, 2006). The DREW diagrams have been demonstrated to support students in reflecting on their previous debate and earlier knowledge (Marttunen and Laurinen, 2007), providing evidence of development of metacognition, a dimension of nonroutine problem solving.

Learning by Design

Kolodner said that, in LBD, students work on a variety of design challenges over the course of the year, using repeated activities. The goal of these activities is to support development of science practices, such as how to design an experiment and how to interpret evidence, that students can apply not only to the immediate design challenge but also to new or nonroutine problems (Kolodner, Gray, and Fasse, 2003). Research on implementation of LBD has identified many anecdotal examples of students applying science practices developed over the course of addressing one design challenge to a new design challenge, as well as spontaneously applying science practices to their science fair projects, without coaching or prompting by the teacher.

The matched comparison study of LBD and non-LBD students described above (Kolodner, Gray, and Fasse, 2003) found that the LBD students scored significantly higher than the comparison group in designing and carrying out an experiment and analyzing the resulting data. These findings suggest that LBD students' skills to solve nonroutine scientific problems are greater than those of non-LBD students.

The LBD curriculum aims to help students develop metacognitive strategies, such as conducting self-checks of their progress when designing an experiment, running an experiment, and analyzing the data. A comparison of average-achieving LBD students with average-achieving students taught using a traditional science curriculum, conducted only two months into the 2000-2001 school year, found that the LBD students scored significantly higher than the comparison group in conducting self-checks. These findings indicate that the LBD curriculum helps students develop metacognitive strategies, an element of nonroutine problem solving, to a greater extent than more traditional science instruction.

Investigating and Questioning our World through Science and Technology

Krajcik observed that the IQWST curriculum engages middle school learners in using evidence and reasoning to build models that describe and explain a host of different phenomena. Studies of implementation of IQWST show that learners improve in their reasoning—specifically, in taking into account sufficient and necessary evidence to support an explanatory model (McNeill and Krajcik, 2008a, 2008b; McNeill et al., 2006). Because scientific reasoning is similar to nonroutine problem solving, these findings suggest that engagement with IQWST increases students' nonroutine problem-solving skills.

Self-Management/Self-Development

Self-management/self-development is defined as the ability to work remotely, in virtual teams; to work autonomously; and to be self-motivating and self-monitoring. One aspect of self-management is the willingness and ability to acquire new information and skills related to work (Houston, 2007).

The 5E Model

Bybee noted that the phases of the 5E instructional model, including the initial "engagement" phase and also the "exploration" phase, in which students explore natural phenomena, are designed to motivate students and increase their interest in science and science learning. He said that studies by Akar (2005), Tinnin (2000), and Von Secker (2002) provide evidence that the 5E model develops student interest in science and positive attitudes toward science learning. Increased interest and positive attitudes may represent development of self-management and self-development.

Online Learning Environments for Argumentation

Clark noted that some of the online environments include participant awareness tools that help students monitor their own contributions and the contributions of other group members, which may encourage self-development/self-management. For example, the Virtual Collaborative Research Institute (VCRI) includes a "shared space," which analyzes chat messages and shows the extent to which group members are conducting shallow online discussions or are engaged in critical exploratory discussion. The tool also visualizes whether group members are agreeing or disagreeing about a topic during online discussion. One study (Jannssen et al., 2007) found that, in comparison to students using VCRI without the shared space tool, students with access to the tool perceived their group's norms and behaviors more positively and their group's strategies as more effective. In addition, students with access to the shared space tool engaged in different collaborative activities, and performed better on one part of a research task in the domain of history. Such participant awareness tools develop students' ability to work in virtual teams, to work autonomously, and to be self-motivating and self-monitoring.

Research suggests that the inclusion of scripts in the CASSIS and WISE environments can promote self-management and self-development (Weinberger et al., 2007). In one study (Wecker and Fischer, 2007), the scripts supporting students in classifying the components of the argumentation of their learning partners and in formulating counterarguments were gradually reduced, and students began to carry out argumentation on their own. These studies indicate that CASSIS develops students' ability to monitor and manage their own performance as well as the performance of others. Research on an early version of WISE (Davis, 2003; Davis and Linn, 2000) showed that generic prompts that ask students to "stop and think" encourage greater reflection in comparison to directed prompts that provide hints indicating potentially productive directions for their reflection. Prompts can support students' self-monitoring, a dimension of self-management/self-development.

Learning by Design

Kolodner noted that the design challenges in LBD require students to identify what skills and concepts they need to learn, carry out investigations to learn what they need to know, and apply their learning. These activities are designed to support students in learning how to monitor and manage their own learning.

The comparison study described above (Kolodner, Gray, and Fasse, 2003) found that LBD students consistently performed significantly better than non-LBD students at conducting self-checks during experiment design, running experiments, and analysis. These findings suggest that engagement with LBD develops students' skills in self-management/self-development of their own learning.

Investigating and Questioning our World through Science and Technology

Krajcik indicated that several different learning activities in IQWST may support development of self-management/self-development. First, students evaluate and critique their own models and scientific explanations as well as those created by others. Studies of IQWST indicate that students improved in these activities (Merritt, Shwartz, and Krajcik, 2008; Shwartz et al., 2008). Second, students use criteria to make judgments, and there is evidence that the curriculum supports improvement in this skill. Third, students consider if they have sufficient and appropriate evidence to support claims, requiring them to monitor and manage their own learning. The studies of IQWST also yield evidence of improvement in these skills (McNeill and Krajcik, 2008a). Finally, adhering to project guidelines and timelines during various projects supports growth in students' self-management and self-development.

Systems Thinking

Systems thinking is defined as the ability to understand how an entire system works; how an action, change, or malfunction in one part of the system affects the rest of the system; and adopting a big-picture perspective on work (Houston, 2007). It includes judgment and decision making, systems analysis, and systems evaluation as well as abstract reasoning about how the different elements of a work process interact (Peterson et al., 1999).

The 5E Model

Bybee observed that understanding of systems thinking may be viewed as a necessary foundation for development and application of systems thinking as a skill. Based on this interpretation, he drew the inference that evidence of the 5E model's effectiveness in enhancing students' mastery of scientific subjects represents evidence of development of systems thinking. Several studies provide "strong" evidence that the model enhances mastery of scientific subject matter, Bybee said (Akar, 2005; Bybee et al., 2006; Coulson, 2002; Taylor et al., 2007; Wilson et al., 2009).

Online Learning Environments for Argumentation

Clark proposed that scientific argumentation and development of systems thinking are related, because arguments are systems and chains of claims, warrants, backings, and data that can involve substantial complexity as they evolve through discussion. In order to participate productively in these discussions, students must learn how to evaluate information, make well-reasoned decisions, and examine how the various components of an argument or counterargument fit together with one another. They must also develop criteria for evaluating what counts as warranted knowledge and how to determine if information is relevant to the phenomenon under discussion, or if there is sufficient information to make a decision. Through these activities, students learn to adopt a big-picture perspective on their work. From this perspective, all of the research showing improvement in argumentation among students engaged with these environments provides evidence of development of systems thinking.

Two studies of DREW, focusing more specifically on systems thinking, suggest that engagement with this environment develops improved understanding of a complex phenomenon or system and supports engagement in systems analysis and evaluation (Marttunen and Laurinen, 2006, 2007).

Learning by Design

Kolodner observed that working toward the design challenges in LBD requires students to develop understanding of a system or set of systems. She proposed that solving such challenges requires judgment and decision making, systems analysis, systems evaluation, and reasoning about how the different elements of a system interact. The curriculum, she said, is designed to support students in developing these dimensions of systems thinking. However, no studies to date have examined development of these dimensions of systems thinking among LBD students.[1]

Investigating and Questioning our World through Science and Technology

Krajcik observed that IQWST supports students' development of systems thinking, but no assessments have been conducted specifically to measure systems thinking. IQWST is designed to enhance student understanding of complex scientific systems, and such understanding may develop systems thinking. The effectiveness of the curriculum in supporting learning of complex scientific content has been studied in large urban areas, suburban areas, and rural areas, including areas with populations of students eligible for free and reduced-price meals. These studies, using pre and posttests, have all shown statistically significant gains in the learning of science concepts (McNeill and Krajcik, 2008a; McNeill et al., 2006; Merritt, Shwartz, and Krajcik, 2008).

Common Themes

Several common themes appear across the promising curriculum models and in other workshop presentations. First, the paper authors' interpretations of the five skills are often quite similar. Clark and Krajcik both view the construction and revision of scientific models, arguments, and explanations as processes that develop adaptability. All of the authors viewed the creation and communication of scientific arguments and explanations—verbally and in writing—as activities that develop complex communication skills. In addition, Clark, Kolodner, and Krajcik identified student work in small groups as supporting development of complex communication skills.

---

[1] Other workshop presenters viewed student gains in understanding of complex scientific systems as evidence of development of systems thinking. Comparative studies of LBD and non-LBD classrooms indicate that LBD students consistently learn science content as well as or better than comparison students, with the largest gains among economically disadvantaged student and students who tested lowest on the pretest (Kolodner et al., 2003).

SYNTHESIS AND REFLECTIONS

Similarly, all of the authors viewed engagement of students in inquiry and development of scientific explanations as supporting development of nonroutine problem solving, and most viewed improvement in students' understanding of complex scientific systems as evidence of development of systems thinking. These authors' shared interpretations of the skills reinforce Schunn's finding that there are many areas of overlap between the science education goals embodied in state and national standards and the five skills.

Second, underlying these common interpretations of the five skills are similar instructional design approaches in the curriculum models. All of the curricula take a problem-based or project-based approach. These curriculum models embed learning of science content (or content in other domains) in investigations and discussions focusing on real-world phenomena or challenges. They are designed to motivate students to learn by engaging them around a driving question, problem, or challenge.

Reinforcing this theme, Anderman and Sinatra recommended that science teachers promote active engagement in inquiry and problem solving, based on connections to adolescent students' personal interests and career goals (see Chapter 3). Windschitl also suggested that the five skills can be developed through problem- or project-based learning, including through scientific inquiry, and proposed that teachers should reconstruct curriculum around a few big ideas in science. Taken together, these presentations suggest that science instruction that embeds student learning in the investigation of real-world problems or phenomena and focuses on a few selected driving questions related to these problems or phenomena is most likely to support development of 21st century skills.

A third theme is that advancing such forms of science instruction would require not only new curriculum designs, but also increased capacity to support teachers. Windschitl called for a continuous improvement system to support teachers in cultivating students' 21st century skills, cautioning that this would require major reforms in science teacher preparation, induction of new teachers, and ongoing professional development. Anderman and Sinatra indicated that adolescents' cognitive capacity to develop the five 21st century skills can be tapped if teachers can motivate them with new teaching and assessment strategies. However, they said that teachers need support from administrators for shared lesson planning and other forms of professional development, as well as training in adolescent development, to design and implement these new strategies. Several other workshop participants highlighted the importance of building capacity to support science teachers in fostering students' 21st century skills.

DISCUSSION GROUP REPORTS

William Sandoval invited all participants to return to the small discussion groups they had been assigned to on the first day. He asked each group to think about and discuss policy options that might support development of 21st century skills in science education and to suggest at least one short-term step that could be taken immediately and one longer term policy option.

Christine Massey (University of Pennsylvania) invited a reporter from each group to share these suggestions. Hanna Doerr (National Commission on Teaching and America's Future) reported that her group's suggestion that, in the short term, the National Council for Accreditation of Teacher Education, the National Association for the Education of Young Children, the Teacher Education Accreditation Council, and other organizations concerned with teacher preparation begin a discussion about integrating 21st century skills into their standards for teacher education and certification. The group's longer term goal is to reform teacher education in order to help future teachers learn 21st century skills and prepare to teach these skills. For example, schools of education could engage future teachers in project-based learning and collaborative study of student assessment results and other student materials.

David Vanier (National Institutes of Health Office of Science Education) reported that, in the short term, his group proposes using some of the money Congress provided to the U.S. Department of Education through the American Recovery and Reinvestment Act of 2009 (commonly known as the stimulus package) to infuse 21st century skills into K-12 education. For example, the funds could be used to support development of education school curriculum materials focused on 21st century skills, to provide grants for teacher education related to 21st century skills, or to forgive student loans to teachers who incorporate 21st century skills. Over the longer term, the group calls for reforming national education standards, both in science and in other school subjects, to incorporate 21st century skills.

The next reporter said that the group proposes, in the short term, to clarify the terms and definitions of 21st century skills, as a common basis for a possible future workshop on assessment of 21st century skills. This group agreed that the skills fall into four categories, including problem solving and critical thinking, flexibility and adaptability, communication skills, and some form of self-direction. Over the longer term, the group advocated development of a research and development agenda for science education that would show how 21st century skills are incorporated into the teaching of science content and provide concrete examples of what these skills look like in curriculum, instruction, and assessment.

Susan Albertine said that her group discussed how the five 21st century skills compared with the goals of science education reform. The members agreed that two of the skills demanded by business—complex communication and nonroutine problem solving—were well aligned with approaches that are gaining support in reform of K-12 and college-level science education, such as problem-based learning, inquiry, and engaging students in design. Albertine said the group proposes, in the short term, to increase clarity about the alignment between 21st century skills and science education reform goals. Over the longer term, this group suggests carrying out a longitudinal study of children to understand what happens over time when they participate in learning environments that emphasize 21st century skills.

Gina Schatteman described her group's short-term goal: to use technology as a leverage point for changing the education system, specifically by using online tools to monitor both student and teacher progress and providing both online and face-to-face mentoring of teachers. The group's long-term goal is to align improved science standards and curriculum to convey 21st century skills, incorporating a progression of learning across grade levels.

The final reporter said the group's immediate action step would be to define the 21st century skills more clearly and operationally. Over the long term, this group thought that universities receiving science research grants from the National Science Foundation should be required to provide expanded research experiences to undergraduates. This suggestion is based on research indicating that such experiences build students' appreciation for both the content and process of science, which may support self-management/self-development (Kardash, 2000).

DISCUSSION

Massey observed that several of the groups asked for clearer definitions of 21st century skills, including their relationship to the goals of science education. She also noted common themes in the areas of incorporating 21st century skills into education standards and assessments and connections between K-12 and higher education, as groups called for changes in colleges of education and in undergraduate science courses, in order to develop teachers' 21st century skills.

Jacob Foster (Massachusetts Department of Education) observed that engaging students in environmentally focused design projects appears to support learning of 21st century skills (Krajcik and Sutherland, 2009; Schunn, 2009). Currently, he said, he must work with three different sets of state education standards, addressing science, technology, and engineering, which can be challenging and confusing. He suggested that, over the long term, it would be valuable to bring together the societies from these three disciplines to develop more integrated standards. As part of this effort, the societies could analyze how the standards relate to, and promote learning of, 21st century skills.

Patricia Harvey (National Research Council) said that members of her group had discussed development of 21st century skills in informal learning environments, such as science museums and field trips. Massey responded that learning outside formal school settings had been mentioned several times and suggested that it would be valuable to involve experts in this field in future activities focusing on 21st century skills.

Ken Kay asked why, when several groups identified selected skills as most relevant to science education, they had dropped systems thinking, a skill he views as an important component of science. Christian Schunn responded that his group left it out because it was unsure whether systems thinking at work differed from scientific analysis of systems. He said this group viewed adaptability and self-management/self-development as elements of high-quality science education, but not as direct goals of science education reform. Based on this view, the group suggested focusing on nonroutine problem solving and complex communication/social skills.

Janet Kolodner said that only one group had narrowed the list of five skills to two (complex communication/social skills and nonroutine problem solving) and that this might be a minority view. She reminded Schunn that he had earlier suggested engaging students in large team projects to develop self-management, indicating that this skill should be a goal of science education. She went on to say that Joseph Krajcik had described students engaged in systems thinking while learning science and that that systems thinking is an essential component of understanding scientific concepts. Although many groups asked for improved definitions of the skills, especially in order to develop assessments, she said, "it would be self-defeating" to assume that all five skills can't be developed in the context of science education.

COMMITTEE MEMBERS' FINAL COMMENTS

In the final workshop session, members of the planning committee offered individual reflections on the intersection of science education and 21st century skills. William Bonvillian began by thanking all participants for sharing their thoughts and insights throughout the workshop. He expressed his view that the commissioned papers had sparked coherent discussions that, in turn, had begun to translate abstract ideas about workforce skills into real actions that could be taken to infuse these skills into science education. He called for further thinking about how best to bring insights from the workshop into science curriculum development, standard setting, teacher professional development, and the creation of technology-enhanced materials.

SYNTHESIS AND REFLECTIONS *101*

Marcia Linn offered special thanks to the industry representatives for helping her and the other science education experts to understand their goals and for listening patiently to the science education experts' different vocabulary to describe skills. She said she was especially impressed by the industry representatives' comments about growing demand for 21st century skills, driven by flat organizations, globalization, and especially by "the fact that everyone changes jobs and responsibilities all the time." Recalling warnings decades earlier about the increasing skill demands of work, Linn said that these warnings have now become reality. She observed that her adult children, like most young adults, change jobs frequently, so they need different skills from those of previous generations.

Linn commented that the theme of systems thinking had emerged frequently, as the workshop participants considered how to meet the increasing demand for 21st century skills in the context of a large, complex education system that is slow to change. She said that technology may offer a leverage point to meet this challenge. She found it puzzling that some education policy leaders argue that schools do not require advanced technology, given the reality that 90 percent of children use technology every day at home, and over 60 percent of middle school and high school students have their own personal websites (Lenhart, Madden, and Hitlin, 2005). She argued that technology skills are important for the 21st century, and students should be able to use and further develop these skills for learning at school.

Linn said that there are many excellent examples of technology-supported science learning environments that encourage students to reflect on and assess their own learning and also allow teachers to view these reflections and respond in real time to student ideas (Linn, 2006; Linn and Eylon, 2006), including the examples discussed at the workshop. Because these systems not only enhance student learning but also provide feedback to teachers for use in changing their instructional practices, Linn said, they offer an important avenue for large-scale change in the education system.

Massey thanked workshop sponsors Ken Kay and Bruce Fuchs as well as the Board on Science Education staff, describing the workshop as "an amazing experience." She said that the workshop had reminded her of the progress being made in science education, including development of good pedagogical models, curriculum materials, and technological learning tools. Research has yielded new insights into design of effective teacher preparation and professional development programs and innovative assessment methods, and these developments can be linked together in more comprehensive science education reform. Massey observed that support is growing for this type of comprehensive reform of science education at all levels, from preschool through the undergraduate level. She suggested that, in thinking about how science education intersects with 21st century skills, it would be valuable to select from, and combine, this array of promising new developments in science education.

Noting that the workshop was designed with an open mind about the extent to which science education might intersect with 21st century skills, Massey said that the presentations and discussions had illuminated many promising intersections. She suggested that communication, collaboration, and systems thinking would be required to further understand and develop these intersections. It is important, she said, to clarify and articulate a synthesized message about the strongest and most promising areas of intersection, framed so that different constituencies (science educators, the business community) can support it and recognize their own interests.

Reflecting on conversations about leverage points in the education system, Massey recalled Mark Windschitl's graphic illustration of the challenge of changing many interrelated components of K-12 and higher education at once, in order to support teaching of 21st century skills (see Figure 6-1). She also recalled Christian Schunn's depiction of many system components, including state and national science standards, state tests, and national organizations, all influencing classroom science teaching (see Figure 2-1). Massey said that workshop participants had begun to identify potential leverage points in the education system, including assessments, science standards, teacher certification requirements, and changes in individual schools. Although there is still a great deal to think about, the workshop provided some depth in moving forward to tackle problems in science education, she concluded.

Carlo Parravano suggested that a consensus study might be valuable to illuminate what good science teaching would look like if it incorporated standards-based science content intertwined with 21st century skills. Such a study could be modeled on *Taking Science to School: Learning and Teaching Science in Grades K-8* (National Research Council, 2007a), which is based on research, and its companion volume, *Ready, Set, Science: Putting Research to Work in K-8 Science Classrooms* (National Research Council, 2008c), which translates that research for practitioners. Parravano said that these two reports have moved the field of science education a great deal in the year and a half since they became available, and a similar effort on science education and 21st century skills would provide a common vision, language, and framework. It would provide guidance for teacher professional development, curriculum and assessment development, and future research needs related to 21st century skills and science education.

In terms of leverage points, Parravano said that parents may be particularly effective. He relayed a favorite observation of a colleague who is closely tied to the policy world—"evidence doesn't move legislators, the public does." To the extent that this is true, he said, further work in this area would help to build a strong vision among the parents and the public about 21st century skills that science education aims to develop. Finally, Parravano said it is important to remember that "the 21st century skills are really a tool and not an end point." A problem with science standards, he said, is that they are now seen as an end point, rather than as a tool to improve science teaching and learning. As a result, they are rewritten in a way that makes their original goals no longer recognizable, and they lose meaning. He suggested working hard to protect the shared understanding of 21st century skills that had developed over the course of the workshop.

William Sandoval added his thanks to all participants and thanked the Board on Science Education staff for giving him the opportunity to work with, and learn from, the other members of the workshop planning committee. He agreed with other committee members that it was valuable to bring people representing different constituencies together to build shared understanding. Sandoval said he believed that the workshop had provided good answers to the six guiding questions developed by the planning committee (see Chapter 1).

Sandoval said he thinks, based on the workshop, that there are extensive and strong areas of overlap between models of high-quality science education and 21st century skills, and that these overlaps represent a positive development. At the same time, however, he cautioned that there are many barriers that stand in the way of systemic change in science education. He encouraged the audience to focus on the promising models of science curriculum, teacher education, and assessment (discussed at the workshop) and promote the expansion of these models in their home communities. He repeated his concern about several important groups that were not represented at the workshop, including parents, scientists who teach undergraduates, and teachers, and suggested including these groups in future conversations. While acknowledging the value of common definitions and shared understandings of 21st century skills, he suggested that these future conversations would also benefit from the "healthy pluralism" that is a hallmark of democracy. Finally, Sandoval explained that he was initially skeptical of the workshop's focus on the skill demands of the economy, including the preliminary definitions of the five skills in job contexts (see Box 1-1). He called for extending the purpose and definitions of these 21st century skills to include civic dispositions and other skills needed to participate effectively in a complex, technologically sophisticated democracy.

Arthur Eisenkraft said that, although he was enriched by the workshop papers and discussions, he does not yet fully understand the meaning of the five 21st century skills. He recalled his search several years ago to identify the top 20 high school physics students in America in order to engage in an international competition. After several rounds of testing, he sent out a very difficult physics problem by mail, but he deliberately left out one crucial ingredient. He did not explicitly mention that the particle was moving at right angles to the field, although the picture he included seemed to show this.

Eisenkraft said he received three different types of written responses on the students' test booklets. Students in one group wrote that, because they did not have complete information, they would not solve the problem. Students in a second group wrote that, lacking complete information, they assumed from the picture that the particle was moving at a 90 degree angle, and they solved the problem. A third, much smaller group, wrote that, although they did not have complete information, they had assumed the 90 degree angle and solved the problem. However, students in this group did not stop there. They explained that, if the particle was not moving at a 90 degree angle, they could not solve the problem, but they nevertheless offered their best guess about whether the solution would change. These were the students he wanted for the competition, Eisenkraft, said, because they had 21st century skills.

He suggested that educators want students to understand when to be adaptable and how adaptable to be. He asked for improved definitions of 21st century skills, with examples of how individuals deploy these skills when carrying out specific tasks at work. It would be helpful, he said, to have a detailed description of the ways in which a janitor with 21st century skills mops a hospital floor to compare with a similarly detailed description of the ways in which a janitor lacking these skills would perform the same task. Such descriptions, he said, could help educators discuss and more clearly define, the skills. They could also help to clarify when 21st century skills should be deployed. Eisenkraft noted the example of Chuck Yeager, a well-known test pilot in the U.S. Air Force. Yeager, he said, rejected the use of instruction manuals describing routine flight procedures and preferred to learn through the actual process of flying. Turning to the field of surgery, Eisenkraft said most surgeries are routine, and the patient prefers the surgeon to follow well-established, successful procedures, rather than being adaptable. Occasionally, however, when something goes wrong, the surgeon's ability to adapt and improvise solutions to nonroutine problems suddenly becomes critical.

Eisenkraft concluded that the workshop was only the first step in a continuing process, with many questions related to the intersection of science education and 21st century skills yet to be answered. He said that the papers had enriched the discussions, and that the active participation of individuals from a variety of constituencies had encouraged all participants to think from different perspectives.

In closing, Ken Kay noted that state education agencies and school districts are at different stages in their efforts to infuse 21st century skills into science and other subjects. Reflecting that the Partnership for 21st Century Skills has established a vision (2003, 2009), he observed that West

Virginia is beginning to implement this vision, and other states still have many questions. Some school districts, he said have developed rubrics to guide teaching and learning of 21st century skills, while others are just beginning to develop such rubrics. Similarly, states and school districts are at different stages of incorporating 21st century skills into teacher professional development. In the context of this continuum of different stages of movement toward 21st century skills, he said, the workshop papers and discussions are very helpful.

References

Abell, S.K., Anderson, G., and Chezem, J. (2000). Science as argument and explanation: Exploring concepts of sound in third grade. In J. Minstrell and E.H. Van Zee (Eds.), *Inquiry into inquiry learning and teaching in science* (pp. 100-119). Washington, DC: American Association for the Advancement of Science.

Akar, E. (2005). *Effectiveness of 5E learning cycle model on students' understanding of acid-base concepts.* Dissertation Abstracts International.

Alonzo, A.C. (2002). *Evaluation of a model for supporting the development of elementary school teachers' science content knowledge.* Proceedings of the Annual International Conference of the Association for the Education of Teachers in Science. Charlotte, NC.

American Association for the Advancement of Science. (1993). *Benchmarks for science literacy.* Washington, DC: Author.

American Association of School Librarians. (2009). *Standards for the 21st-century learner.* Available: http://www.ala.org/ala/mgrps/divs/aasl/guidelinesandstandards/learningstandards/standards.cfm [retrieved June 18, 2009].

Anderman, E.M., and Anderman, L.H. (2009). *Classroom motivation.* Boston: Pearson.

Anderman, E.M., and Sinatra, G.M. (2009). *The challenges of teaching and learning about science in the 21st century: Exploring the abilities and constraints of adolescent learners.* Paper prepared for the Workshop on Exploring the Intersection of Science Education and the Development of 21st Century Skills, National Research Council. Available: http://www7.nationalacademies.org/bose/AndermanSinatra.pdf [retrieved May 2009].

Anderman, E.M., Eccles, J.S., Yoon, K.S., Roeser, R.W., Wigfield, A., and Blumenfeld, P. (2001). Learning to value math and reading: Individual differences and classroom effects. *Contemporary Educational Psychology, 26,* 76-95.

Anderman, E.M., Griessinger, T., and Westerfield, G. (1998). Motivation and cheating during early adolescence. *Journal of Educational Psychology, 90,* 84-93.

Anderson, J.R. (1983). A spreading activation theory of memory. *Journal of Verbal Learning and Verbal Behavior, 22,* 261-295.

REFERENCES

Andrew, J.P., DeRocco, E.S., and Taylor, A. (2009). *The innovation imperative in manufacturing: How the United States can restore its edge.* Boston: Boston Consulting Group, Inc. Available: http://www.nam.org/~/media/AboutUs/ManufacturingInstitute/innovationreport.ashx [retrieved March 2009].

Atkin, J.M., and Karplus, R. (1962). Discovery or invention? *The Science Teacher, 29,* 45-51.

Autor, D.H., Levy, F., and Murnane, R.J. (2003). The skill content of recent technological change: An empirical exploration. *Quarterly Journal of Economics, 118*(4), 1279-1333.

Baldi, S., Jin, Y., Skemer, M., Green, P., Herget, D., and Xie, H. (2007). *Highlights from PISA 2006: Performance of U.S. 15-year old students in science and mathematics literacy in an international context.* Washington, DC: National Center for Education Statistics, U.S. Department of Education.

Banilower, E., Cohen, K., Pasley, J. and Weiss, I. (2008). Effective science instruction: What does research tell us? Portsmouth, NH: RMC Research Corporation, Center on Instruction. Available: http://www.centeroninstruction.org/files/Characteristics%20of%20Effective%20Science%20Instruction%20REVISED%20FINAL.pdf [retrieved September 2009].

Barrows, H.S. (1985). *How to design a problem-based curriculum for the preclinical years.* New York: Springer.

Bell, P., and Linn, M.C. (2000). Scientific arguments as learning artifacts: Designing for learning from the web with KIE. *International Journal of Science Education, 22*(8), 797-817.

Bennett, R.E., Persky, H., Weiss, A.R., and Jenkins, F. (2007). *Problem-solving in technology-rich environments. A report from the NAEP Technology-Based Assessment Project.* Research and Development Series. Institute of Education Sciences, NCES 2007-466. Washington, DC: U.S. Department of Education. Available: http://nces.ed.gov/nationsreportcard/pdf/studies/2007466_1.pdf [retrieved June 23, 2009].

Bloom, B.S. (1956). *Taxonomy of educational objectives, handbook I: The cognitive domain.* New York: David McKay.

Blumenfeld, P., Soloway, E., Marx, R.W., Guzdial, M., and Palincsar, A. (1991). Motivating project-based learning: Sustaining the doing, supporting the learning. *Educational Psychologist, 26*(3/4), 369-398.

Boddy, M., Watson, K., and Aubusson, P. (2003). A trial of the five Es: A referent model for constructivist teaching and learning. *Research in Science Education, 33*(1), 27-42.

Bransford, J.D., and Schwartz, D.L. (1999). Rethinking transfer: A simple proposal with multiple implications. *Review of Research in Education, 24*(1), 61-100.

Brickhouse, N.W. (1990). Teacher beliefs about the nature of science and their relationship to classroom practices. *Journal of Teacher Education, 41*(3), 53-62.

Bybee, R.W. (2009). *The BSCS 5E instructional model and 21st century skills.* Paper prepared for the Workshop on Exploring the Intersection of Science Education and the Development of 21st Century Skills, National Research Council. Available: http://www7.nationalacademies.org/bose/21CentSKillUploads.html [retrieved May 2009].

Bybee, R., Taylor, J., Gardner, A., Van Scotter, P., Powell, J., Westbrook, A., and Landes, N. (2006). *The BSCS 5E instructional model: Origins and effectiveness.* Colorado Springs, CO: BSCS.

Cacioppo, J.T., Petty, R.E, Feinstein, J.A., and Jarvis, W.B.G. (1996). Dispositional differences in cognitive motivation: The life and times of individuals varying in need for cognition. *Psychological Bulletin, 119*(2), 197-253.

Case, E., Stevens, R., and Cooper, M. (2007). Is collaborative grouping an effective instructional strategy? *Journal of College Science Teaching, 36*(6), 42-47.

Casner-Lotto, J., and Barrington, L. (2006). *Are they really ready to work?* Washington, DC: Conference Board, Partnership for 21st Century Skills, Corporate Voices for Working Families, and Society for Human Resource Management. Available: http://www.conference-board.org/Publications/describe.cfm?id=1218 [retrieved March 2009].

Chase, W.G., and Simon, H.A. (1973). Perception in chess. *Cognitive Psychology, 4,* 55-81.

Chi, M.T.H. (2005). Common sense conceptions of emergent processes: Why some misconceptions are robust. *Journal of the Learning Sciences, 14,* 161-199.

Chi, M.T.H., and Koeske, R.D. (1983). Network representation of a child's dinosaur knowledge. *Developmental Psychology, 19,* 29-39.

Clark, D.B. (2004). Hands-on investigation in Internet environments: Teaching thermal equilibrium. In M.C. Linn, E.A. Davis, and P. Bell (Eds.), *Internet environments for science education* (pp. 175-200). Mahwah, NJ: Lawrence Erlbaum Associates.

Clark, D.B., and Sampson, V.D. (2005, June). *Analyzing the quality of argumentation supported by personally-seeded discussions.* Paper presented at the annual meeting of the Computer Supported Collaborative Learning (CSCL) Conference, Taipei, Taiwan.

Clark, D.B., and Sampson, V.D. (2006, July). *Evaluating argumentation in science: New assessment tools.* Paper presented at the International Conference of the Learning Sciences, Bloomington, Indiana.

Clark, D.B., and Sampson, V.D. (2008). Assessing dialogic argumentation in online environments to relate structure, grounds, and conceptual quality. *Journal of Research in Science Teaching, 45*(3), 6.

Clark, D.B., D'Angelo, C.M., and Menekse, M. (in press). Initial structuring of online discussions to improve learning and argumentation: Incorporating students' own explanations as seed comments versus an augmented-preset approach to seeding discussions. Submitted to the *Journal of Science Education and Technology.*

Clark, D.B., Menekse, M., D'Angelo, C., Touchman, S., and Schleigh, S. (2008). *Scaffolding students' argumentation about simulations.* Paper presented as part of a symposium organized by Hsin-Yi Chang to the International Conference of the Learning Sciences (ICLS) 2008, Utrecht, Netherlands.

Clark, D.B., Sampson, V., Stegmann, K., Marttunen, M., Kollar, I., Janssen, J., Weinberger, A., Menekse, M., Erkens, G., and Laurinen, L. (2009). *Scaffolding scientific argumentation between multiple students in online learning environments to support the development of 21st century skills.* Paper prepared for the Workshop on Exploring the Intersection of Science Education and the Development of 21st Century Skills, National Research Council. Available: http://www7.nationalacademies.org/bose/21CentSKillUploads.html [retrieved May 2009].

Cooper, M.M., Sandi-Urena, S., and Stevens, R. (2008). Reliable multi-method assessment of metacognition use in chemistry problem-solving. *Chemical Education Research and Practice, 9,* 18-24.

Cooper, M.M., Stevens, R., and Holme, T. (2006). Assessing problem-solving in chemistry using the IMMEX system. *Proceedings of the National STEM Assessment Conference* (pp. 118-129).

Coulson, D. (2002). *BSCS Science: An inquiry approach—2002 evaluation findings.* Arnold, MD: PS International.

Cuthbert, A.J., Clark, D.B., and Linn, M.C. (2002). WISE learning communities: Design considerations. In K.A. Renninger and W. Shumar (Eds.), *Building virtual communities: Learning and change in cyberspace* (pp. 215-246). Cambridge: Cambridge University Press.

Darling-Hammond, L. (1999). Target time toward teachers. *Journal of Staff Development, 20,* 31-36.

REFERENCES

Darling-Hammond, L., and Cobb, V.L (Eds.). (1995). *Teacher preparation and professional development in APEC members: A comparative study.* ED 383 683. Washington, DC: U.S. Department of Education.

Darr, A. (2007). *The knowledge worker and the future skill demands of the U.S. workforce.* Paper prepared for the Workshop on Research Evidence Related to Future Skill Demands, National Research Council. Available: http://www7.nationalacademies.org/cfe/Future_Skill_Demands_Asaf_Darr_Paper.pdf [retrieved March 2009].

Davis, E.A. (2003). Prompting middle school science students for productive reflection: Generic and directed prompts. *Journal of the Learning Sciences, 12*(1), 91-142.

Davis, E.A., and Linn, M.C. (2000). Scaffolding students' knowledge integration: Prompts for reflection in KIE. *International Journal of Science Education, 22*(8), 819-837.

Deloitte Development, LLC, and the Manufacturing Institute. (2005). *2005 skills gap report—A survey of the American workforce.* Washington, DC: Deloitte Development, LLC. Available: http://www.nam.org/~/media/AboutUs/ManufacturingInstitute/innovationreport.ashx [retrieved March 2009].

DeSimone, L.M., Porter, A.S., Garet, M.S., Yoon, K.S., and Birman, B. (2002). Effects of professional development on teachers' instruction: Results from a three-year longitudinal study. *Educational Evaluation and Policy Analysis, 24*(2), 81-112.

Dole, J.A., and Sinatra, G.M. (1998). Reconceptualizing change in the cognitive construction of knowledge. *Educational Psychologist, 33*(2/3), 109-128.

Fosnot, C.T. (1996). Constructivism: A psychological theory of learning. In C.T. Fosnot (Ed.), *Constructivism: Theory, perspectives, and practice* (pp. 8-33). New York: Teachers College Press.

Franklin, J.C. (2007). Employment outlook: 2006-16: An overview of BLS projections to 2016. *Monthly Labor Review*, November. Available: http://www.bls.gov/opub/mlr/2007/11/art1abs.htm [retrieved March 2009].

Garet, M.S., Porter, A.C., Desimone, L., Birman, B.F., and Yoon, K.S. (2001). What makes professional development effective? Results from a national sample of teachers. *American Educational Research Journal, 38*(4), 915-945.

Gatta, M., Boushey, H., and Appelbaum, E. (2007). *High-touch and here-to-stay: Future skill demands in low wage service occupations.* Paper prepared for the Workshop on Research Evidence Related to Future Skill Demands, National Research Council. Available: http://www7.nationalacademies.org/cfe/Future_Skill_Demands_Mary_Gatta_Paper.pdf [retrieved September 2009].

Gess-Newsome, J., and Lederman, N. (1993). Pre-service biology teachers' knowledge structures as a function of professional teacher education: A year-long assessment. *Science Education, 77*(1), 25-46.

Gobet, F., and Simon, H.A. (1996). Recall of random and distorted chess positions: Implications for the theory of expertise. *Memory and Cognition, 24*(4), 493-503.

Graber, K.C. (1996). Influencing student beliefs: The design of a "high impact" teacher education program. *Teaching and Teacher Education, 12*, 451-466.

Greene, J.A., and Azevedo, R. (2007). Adolescents' use of self-regulatory processes and their relation to qualitative mental model shifts while using hypermedia. *Journal of Educational Computing Research, 36*, 125-148.

Guzdial, M., and Turns, J. (2000). Effective discussion through a computer-mediated anchored forum. *Journal of the Learning Sciences, 9*(4), 437-470.

Houston, J. (2007). *Future skill demands, from a corporate consultant perspective.* Presentation at the Workshop on Research Evidence Related to Future Skill Demands, National Research Council. Available: http://www7.nationalacademies.org/cfe/Future_Skill_Demands_Presentations.html [retrieved March 2009].

Houston, J.S., and Cochran, C.C. (2009). *Corporate assessment of 21st century skills*. Presentation at the Workshop on Exploring the Intersection of Science Education and the Development of 21st Century Skills, National Research Council. Available: http://www7.nationalacademies.org/bose/Houston_21st_Century_Presentation.pdf [retrieved September 2009].

Inhelder, B., and Piaget, J. (1958). *The growth of logical thinking from childhood to adolescence*. New York: Basic Books.

Iowa Area Education Agencies. (2005). *Science standard 1. Interval benchmark, Grade 10*. Available: http://www.integratingstandards.com/standards/science/standard1_benchmark3_grade10.html [retrieved April 2009].

Jackson, S., Stratford, J., Krajcik, S., and Soloway, E. (1996). Making system dynamics modeling accessible to pre-college science students. *Interactive Learning Environments*, 4, 233-257.

Janssen, J., Erkens, G., and Kanselaar, G. (2007). Visualization of agreement and discussion processes during computer-supported collaborative learning. *Computers in Human Behavior*, 23, 1105-1125.

Janssen, J., Erkens, G., Kanselaar, G., and Jaspers, J. (2007). Visualization of participation: Does it contribute to successful computer-supported collaborative learning? *Computers and Education*, 49, 1037-1065.

Kali, Y., Linn, M.C., and Roseman, J.E. (2008). *Designing coherent science education: Implications for curriculum, instruction, and policy*. New York: Teachers College Press.

Kansas Department of Education. (2007). *Science standards*. Available: http://www.ksde.org/Default.aspx?tabid=144 [retrieved April 2009].

Kardash, C.M. (2000). Evaluation of an undergraduate research experience: Perceptions of undergraduate interns and their faculty mentors. *Journal of Educational Psychology*, 92(1), 191-201.

Karplus, R., and Their, H.D. (1967). *A new look at elementary school science*. Chicago: Rand McNally.

Klahr, D., and Carver, S.M. (1988). Cognitive objectives in a logo debugging curriculum: Instruction, learning, and transfer. *Cognitive Psychology*, 20, 362-404.

Klahr, D., and Nigam, M. (2004). The equivalence of learning paths in early science instruction: Effects of direct instruction and discovery learning. *Psychological Science*, 15(10), 661-667.

Klein, S.P., Freedman, D., Shavelson, R.J., and Bolus, R. (in press). Assessing school effectiveness. *Evaluation Review*.

Kolodner, J.L. (1993). *Case-based reasoning*. San Mateo, CA: Morgan Kaufmann.

Kolodner, J.L. (2009). *Learning by Design's framework for promoting learning of 21st century skills*. Presentation to the Workshop on Exploring the Intersection of Science Education and the Development of 21st Century Skills, National Research Council. Available: http://www7.nationalacademies.org/bose/Kolodner.pdf [retrieved June 2009].

Kolodner, J.L., Camp, P.J., Crismond, D., Fasse, B.B., Gray, J., Holbrook, J., Puntambekar, S., and Ryan, M. (2003). Problem-based learning meetings case-based reasoning in the middle school science classroom: Putting Learning by Design into practice. *Journal of the Learning Sciences*, 12(4), 495-547.

Kolodner, J.L., Gray, J., and Fasse, B.B. (2003). Promoting transfer through case-based reasoning: Rituals and practices in learning by design classrooms. *Cognitive Science Quarterly*, 3(2), 183-232.

Koschmann, T.D., Myers, A.C., Feltovich, P.J., and Barrows, H.S. (1994). Using technology to assist in realizing effective learning and instruction: A principled approach to the use of computers in collaborative learning. *Journal of the Learning Sciences*, 3, 225-262.

REFERENCES

Krajcik, J., and Blumenfeld, P.C. (2006). Project-based learning. In R K. Sawyer (Ed.), *The Cambridge handbook of the learning sciences* (pp. 333-354). New York: Cambridge University Press.

Krajcik, J.S., and Sutherland, L. (2009). *IQWST materials: Meeting the challenges of the 21st century.* Paper prepared for the Workshop on Exploring the Intersection of Science Education and the Development of 21st Century Skills, National Research Council. Available: http://www7.nationalacademies.org/bose/Krajcik_Sutherland_Comm%20Paper.pdf [retrieved May 2009].

Krajcik, J., McNeill, K.L., and Reiser, B. (2008). Learning-goals-driven design model: Developing curriculum materials that align with national standards and incorporate project-based pedagogy. *Science Education*, 92(1), 1-32.

Krajcik, J., Slotta, J., McNeill, K.L., and Reiser, B (2008). Designing learning environments to support students constructing coherent understandings. In Y. Kali, M.C. Linn, and J.E. Roseman (Eds.), *Designing coherent science education: Implications for curriculum, instruction, and policy.* New York: Teachers College Press.

Kuhn, D., and Phelps, E. (1982). The development of problem-solving strategies. In H. Reese (Ed.), *Advances in child development and behavior* (vol. 17, pp. 1-44). New York: Academic Press.

Kuhn, L., and Reiser, B. (2005). *Students constructing and defending evidence-based scientific explanations.* Paper presented at the annual meeting of the National Association for Research in Science Teaching, Dallas.

Lave, J. and Wenger, E. (1991). *Situated learning: Legitimate peripheral participation.* Cambridge, UK: Cambridge University Press.

Lederman, N.G. (1999). Teachers' understanding of the nature of science and classroom practice: Factors that facilitate or impede the relationship. *Journal of Research in Science Teaching*, 36(8), 916-929.

Lemke, C., Coughlin, E., Thadani, V., and Martin, C. (2003). *EnGauge 21st century skills for 21st century learners: Literacy in the digital age.* Naperville, IL: North Central Regional Education Laboratory and the Metiri Group. Available: http://www.metiri.com/features.html [retrieved June 18, 2009].

Lemke, M., Sen, A., Pahlke, E., Partelow, L., Miller, D., Williams, T., Kastberg, D., and Jocelyn, L. (2004). *International outcomes of learning in mathematics literacy and problem solving: PISA 2003 Results from the U.S. perspective.* Washington, DC: National Center for Education Statistics (NCES-2005003). Available: http://nces.ed.gov/pubsearch/pubsinfo.asp?pubid=2005003 [retrieved January 2005].

Lenhart, A., Madden, M., and Hitlin, P. (2005, July 27). *Teens and technology: Youth are leading the transition to a fully wired and mobile nation.* Washington, DC: Pew Internet and American Life Project. Available from http://www.pewinternet.org/Reports/2005/Teens-and-Technology.aspx [retrieved April 2009].

Levy, F., and Murnane, R.J. (2004). *The new division of labor: How computers are creating the next job market.* Princeton, NJ: Princeton University Press.

Lewis, C., and Tsuchida, I. (1997). Planned educational change in Japan: The case of elementary science instruction. *Journal of Educational Policy*, 12(5), 313-331.

Li, J., Klahr, D., and Siler, S.A. (2006). What lies beneath the science achievement gap: The challenges of aligning science instruction with standards and tests. *Science Educator*, 15(1), 1-12.

Li, M. (2001). *A framework for science achievement and its link to test items.* Unpublished dissertation. Stanford, CA: Stanford University.

Li, M., and Tsai, S. (2007). *Linking assessment to science achievement.* NSF Technical report. [See http://www.nsf.gov/awardsearch/showAward.do?AwardNumber=0310360 for description of the research project.]

Li, M., Ruiz-Primo, M.A., and Shavelson, R.J. (2006). Towards a science achievement framework: The case of TIMSS 1999. In S. Howie and T. Plomp (Eds.), *Contexts of learning mathematics and science: Lessons learned from TIMSS* (pp. 291-311). London: Routledge.

Linn, M. (2006). The knowledge integration perspective on learning and instruction. In R.K. Sawyer (Ed.), *The Cambridge handbook of the learning sciences* (pp. 243-264). New York: Cambridge University Press.

Linn, M.C., and Eylon, B.S. (2006). Science education: Integrating views of learning and instruction. In P.A. Alexander and P.H. Winne (Eds.), *Handbook of educational psychology*, 2nd ed. (pp. 511-544). Mahwah, NJ: Lawrence Erlbaum Associates.

Linn, M., Davis, E. and Bell, P. (Eds). (2004). *Internet environments for science education*. Mahwah, NJ: Lawrence Erlbaum Associates.

Loucks-Horsley, S., Hewson, P.W., Love, N., and Stiles, K.E. (1998). *Designing professional development for teachers of science and mathematics*. Thousand Oaks, CA: Corwin Press.

Ma, L. (1999). *Knowing and teaching elementary mathematics*. Mahwah, NJ: Lawrence Erlbaum Associates.

Maehr, M.L., and Midgley, C. (1996). *Transforming school cultures*. Boulder, CO: Westview Press.

Magnusson, S., Borko, H., Krajcik, J.S., and Layman, J.W. (1992). *The relationship between teacher content and pedagogical content knowledge and student content knowledge of heat energy and temperature*. Paper presented at the annual meeting of the National American Association for Research in Science Teaching, Boston, MA.

Maine Department of Education. (2007). *Chapter 132—Learning results: Parameters for essential instruction*. Available: http://www.maine.gov/education/lres/pei/ch132_0708.pdf [retrieved April 2009].

Mäkitalo, K., Weinberger, A., Häkkinen, P., Järvelä, S., and Fischer, F. (2005). Epistemic cooperation scripts in online learning environments: Fostering learning by reducing uncertainty in discourse? *Computers in Human Behavior, 21*(4), 603-622.

Marton, F., and Tsui, A.B.M. (2004). *Classroom discourse and the space of learning*. Mahwah, NJ: Lawrence Erlbaum Associates.

Marttunen, M., and Laurinen, L. (2006). Collaborative learning through argument visualisation in secondary school. In S.N. Hogan (Ed.), *Trends in learning research* (pp. 119-138). New York: Nova Science.

Marttunen, M., and Laurinen, L. (2007). Collaborative learning through chat discussions and argument diagrams in secondary school. *Journal of Research on Technology in Education, 40*(1), 109-126.

Massachusetts Department of Education. (2006). *Massachusetts science and technology/engineering curriculum framework*. Boston: Author. Available: http://www.doe.mass.edu/frameworks/scitech/1006.pdf [retrieved April 2009].

Mathews, J. (2009, January 5). The rush for "21st-century skills": New buzz phrase draws mixed interpretations from educators. *The Washington Post*, p. B2.

Maxwell, N.L. (2006). *The working life: The labor market for workers in low-skilled jobs*. Kalamazoo, MI: W.E. Upjohn Institute for Employment Research.

McNeill, K.L., and Krajcik, J. (2008a). Middle school students' use of appropriate and inappropriate evidence in writing scientific explanations. In M. Lovett and P. Shah (Eds.), *Thinking with data: The proceedings of 33rd Carnegie Symposium on Cognition*. Mahwah, NJ: Lawrence Erlbaum Associates.

McNeill, K.L., and Krajcik, J. (2008b). Scientific explanations: Characterizing and evaluating the effects of teachers' instructional practices on student learning. *Journal of Research in Science Teaching, 45*(1), 53-78.

REFERENCES

McNeill, K.L., Lizotte, D.J., Krajcik, J., and Marx, R.W. (2006). Supporting students' construction of scientific explanations by fading scaffolds in instructional materials. *The Journal of the Learning Sciences*, 15(2), 153-191.

Merritt, J., Shwartz, Y., and Krajcik, J. (2008). Middle school students' development of the particle model of matter. In *Proceedings of the International Conference of the Learning Sciences*, Utrecht, Netherlands.

Murnane, R.J., and Levy, F. (1996). *Teaching the new basic skills: Principles for educating children to thrive in a changing economy.* New York: Free Press and Simon and Schuster.

National Research Council. (1984). *High schools and the changing workplace: The employers' view.* Washington, DC: National Academy Press.

National Research Council. (1996). *National science education standards.* Washington, DC: National Academy Press.

National Research Council. (1999). *How people learn: Brain, mind, experience and school.* Washington, DC: National Academy Press.

National Research Council. (2000). *How people learn: Brain, mind, experience and school: Expanded edition.* Washington, DC: National Academy Press.

National Research Council. (2002). *Learning and understanding: Improving advanced study of mathematics and science in U.S. high schools.* Washington, DC: The National Academies Press.

National Research Council. (2005). *America's lab report: Investigations in high school science.* Washington, DC: The National Academies Press.

National Research Council. (2007a). *Taking science to school: Learning and teaching science in grades K-8.* Washington, DC: The National Academies Press.

National Research Council. (2007b). *Enhancing professional development for teachers: Potential uses of information technology.* Washington, DC: The National Academies Press.

National Research Council. (2008a). *Research on future skill demands: A workshop summary.* Margaret Hilton, Rapporteur. Washington, DC: The National Academies Press.

National Research Council. (2008b). *Common standards for K-12 education? Considering the evidence: Summary of a workshop series.* Alexandra Beatty, Rapporteur. Washington, DC: The National Academies Press.

National Research Council. (2008c). *Ready, set, science: Putting research to work in K-8 science classrooms.* Washington, DC: The National Academies Press.

New Jersey Department of Education. (2004). *New Jersey core curriculum content standards.* Trenton: Author. Available from http://www.nj.gov/education/cccs/cccs.pdf [retrieved April 2009].

North Carolina Public Schools. (2004). *Science standard course of instruction and grade-level competencies.* Raleigh: Author. Available: http://www.dpi.state.nc.us/docs/curriculum/science/scos/2004/science.pdf [retrieved April 2009].

Ogle, D.S. (1986). K-W-L group instructional strategy. In A.S. Palincsar, D.S. Ogle, B.F. Jones, and E.G. Carr (Eds.), *Teaching reading as thinking* (Teleconference Resource Guide, pp. 11-17). Alexandria, VA: Association for Supervision and Curriculum Development.

Ohlsson, S. (1992). The cognitive skill of theory articulation: A neglected aspect of science education? *Science and Education*, 1, 181-192.

Organisation for Economic Co-operation and Development. (2006). *Assessing scientific literacy, reading, and mathematical literacy: A framework for PISA 2006* (p. 34). Available: http://www.oecd.org/dataoecd/63/35/37464175.pdf [retrieved June 24, 2009].

Paris, S.G., Lipson, M.Y., and Wixson, K.K. (1983). Becoming a strategic reader. *Contemporary Educational Psychology*, 8, 293-316.

Partnership for 21st Century Skills. (2003). *The road to 21st century learning: A policymakers' guide to 21st century skills.* Washington, DC: Author. Available: http://www.21stcenturyskills.org/index.php?option=com_contentandtask=viewandid=30and Itemid=185 [retrieved June 2009].

Partnership for 21st Century Skills. (2009a). *State initiatives: Overview of state leadership initiative.* Available: http://www.21stcenturyskills.org/index.php?option=com_content andtask=viewandid=505andItemid=189 [retrieved June 2009].

Partnership for 21st Century Skills. (2009b). *21st century learning environments.* Tucson, AZ: Author. Available: http://www.21stcenturyskills.org/documents/le_white_paper-1.pdf [retrieved April 2009].

Peterson, N., Mumford, M., Borman, W., Jeanneret, P., and Fleishman, E. (1999). *An occupational information system for the 21st century: The development of O*NET.* Washington, DC: American Psychological Association.

Pintrich, P.R. (2000). The role of goal orientation in self-regulated learning. In M. Boekaerts, P.R. Pintrich, and M. Zeidner (Eds.), *Handbook of self-regulation* (pp. 451-502). San Diego: Academic.

Pintrich, P.R., and Schunk, D.H. (2002). *Motivation in education: Theory, research, and applications*, 2nd ed. Upper Saddle River, NJ: Merrill/Prentice Hall.

Pulakos, E.D., Arad, S., Donovan, M.A., and Plamondon, K.E. (2000). Adaptability in the workplace: Development of a taxonomy of adaptive performance. *Journal of Applied Psychology, 85,* 612-624.

Putnam, R.T., and Borko, H. (2000). What do new views of knowledge and thinking have to say about research on teacher learning? *Educational Researcher, 29*(1), 4-15.

Ravitch, D. (2009). *21st century skills: An old familiar song.* Washington, DC: Common Core. Available: http://www.commoncore.org/pressreleases.php [retrieved March 2, 2009].

Roehrig, G., and Luft, J. (2004). Constraints experienced by beginning secondary science teachers in implementing scientific inquiry lessons. Research Report. *International Journal of Science Education, 26*(1), 3-24.

Roth, K.J, and Garnier, H.E. (2006/2007). What science teaching looks like: An international perspective. *Educational Leadership, 64*(4), 16-23.

Rotherham, A. (2008, December 15). 21st-century skills are not a new education trend, but could be a fad. *U.S. News and World Report.* Available : http://www.usnews.com/articles/opinion/2008/12/15/21st-century-skills-are-not-a-new-education-trend-but-could-be-a-fad.html [retrieved March 2009].

Ruiz-Primo, M.A. (1997). *Toward a framework of subject-matter achievement assessment.* Unpublished manuscript. Stanford, CA: Stanford University.

Ruiz-Primo, M.A. (1998). *Models for measuring science achievement.* Invited talk. National Center for Research on Evaluation, Standards, and Student Testing Conference, University of California, Los Angeles.

Ruiz-Primo, M.A. (2003). *A framework to examine cognitive validity.* Paper presented at the meeting of the American Education Research Association, Chicago.

Ruiz-Primo, M.A. (2009). *Towards a framework for assessing 21st century science skills.* Paper prepared for the Workshop on Exploring the Intersection of Science Education and the Development of 21st Century Skills, National Research Council. Available: http://www7.nationalacademies.org/bose/RuizPrimo.pdf [retrieved June 2009].

Salminen, T., Marttunen, M., and Laurinen, L. (2007). Collaborative argument diagrams based on dyadic computer chat discussions. In R. Kinshuk, D.G., Sampson, J.M., Spector and P. Isaias (Eds.), *Proceedings of the IADIS international conference on cognition and exploratory learning in the digital age* (pp. 197-204). December 7-9, Algarve, Portugal.

Sanders, L.R., Borko, H., and Lockard, J.D. (1993). Secondary science teachers' knowledge base when teaching science courses in and out of their area of certification. *Journal of Research in Science Teaching, 30*(7), 723-736.

Sandoval, W.A. (2003). Conceptual and epistemic aspects of students' scientific explanations. *Journal of the Learning Sciences, 12*(1), 5-51.

Sawchuk, S. (2009, January 7). "21st century skills" focus shifts teachers' role. *Education Week*.

Schank, R.C. (1982). *Dynamic memory.* New York: Cambridge University Press.

Schank, R.C. (1999). *Dynamic memory revisited.* New York: Cambridge University Press.

Schank, R.C., and Abelson, R.P. (1977). *Scripts, plans, goals, and understanding: An inquiry into human knowledge structures.* Hillsdale, NJ: Lawrence Erlbaum Associates.

Schmidt, W.H., Wang, H.C., and McKnight, C.C. (2005). Curriculum coherence: An examination of U.S. mathematics and science content standards from an international perspective. *Journal of Curriculum Studies, 37*(5), 525-559.

Schunk, D.H., and Ertmer, P.A. (1999). Self regulatory processes during computer skill acquisition: Goal and self-evaluative influences. *Journal of Educational Psychology, 91,* 251-260.

Schunk, D.S., and Zimmerman, B.J. (2008). *Motivation and self-regulated learning: Theory, research, and applications.* Mahwah, NJ: Lawrence Erlbaum Associates.

Schunn, C. (2009). *Are 21st century skills found in science standards?* Paper prepared for the Workshop on Exploring the Intersection of Science Education and the Development of 21st Century Skills, National Research Council. Available: http://www7.nationalacademies.org/bose/Schunn.pdf [retrieved March 2009].

Schwarz, C.V., Reiser, P., Davis, E.A., Kenyon, L.O., Acher, A., Fortus, D., Schwartz, Y., Hug, B., and Krajcik, J. (2009). Developing a learning progression of scientific modeling: Making scientific modeling accessible and meaningful for learners. *Journal of Research in Science Teaching, 46*(6), 632-654.

Shavelson, R.J., Ruiz-Primo, M.A., Li, M., and Ayala, C.C. (2002). *Evaluating new approaches to assessing learning.* CSE Technical Report 604. Los Angeles: Center for Research on Evaluation, Standards, and Student Testing, University of California, Los Angeles.

Shwartz, Y., Weizman, A., Fortus, D., Krajcik, J., and Reiser, B. (2008). The IQWST experience: Using coherence as a design principle for a middle school science curriculum. *The Elementary School Journal, 109*(2), 199-219.

Singley, M.K., and Anderson, J.R. (1989). *The transfer of cognitive skill.* Cambridge, MA: Harvard University Press.

Slavin, R.E. (1995). *Cooperative learning,* 2nd ed. Boston: Allyn and Bacon.

State of Minnesota. (2009). *Competency modeling clearinghouse.* St. Paul: Author. Available: http://www.careeronestop.org/CompetencyModel/learnCM.aspx [retrieved March 2009].

Stegmann, K., Weinberger, A., and Fischer, F. (2007). Facilitating argumentative knowledge construction with computer-supported collaboration scripts. *International Journal of Computer-Supported Collaborative Learning, 2*(4), 421-447.

Stegmann, K., Wecker, C., Weinberger, A., and Fischer, F. (2007). Collaborative argumentation and cognitive processing: An empirical study in a computer-supported collaborative learning environment. In C. Chinn, G. Erkens, and S. Puntambekar (Eds.), *Mice, minds, and society* (pp. 661-670). New Brunswick, NJ: International Society of the Learning Sciences.

Taylor, J., Van Scotter, P., and Coulson, D. (2007). Bridging research on learning and student achievement: The role of instructional materials. *Science Educator, 16*(2), 44-50.

Thorndike, E.L., and Woodworth, R.S. (1901). The influence of improvement in one mental function upon the efficiency of other functions. *Psychological Review, 8,* 247-261.

Tinnin, R. (2000). The effectiveness of a long-term professional development program on teachers' self-efficacy, attitudes, skills, and knowledge using a thematic learning approach. *Dissertation Abstracts International*, 61(11), 4345.

Tomlinson, C.A. (2003). *Fulfilling the promise of the differentiated classroom: Strategies and tools for responsive teaching.* Alexandria, VA: Association for Supervision and Curriculum Development.

U.S. Department of Labor. (1991). *What work requires of schools: A SCANS report for America 2000.* Secretary's Commission on Achieving Necessary Skills. Available: http://wdr.doleta.gov/SCANS/whatwork/whatwork.pdf [retrieved May 2009].

Vogel, G. (2007). Science education: Global review faults U.S. curricula. *Science*, 274(5286), 335.

Von Secker, C. (2002). Effects of inquiry-based teacher practices on science excellence and equity. *Journal of Educational Research*, 95, 151-160.

Wecker, C., and Fischer, F. (2007). Fading scripts in computer-supported collaborative learning: The role of distributed monitoring. In C. Chinn, G. Erkens, and S. Puntambekar (Eds.), *Mice, minds, and society* (S. 763-771). New Brunswick, NJ: International Society of the Learning Sciences.

Weinberger, A. (2008). *CSCL scripts: Effects of social and epistemic scripts on computer-supported collaborative learning.* Berlin: VDM Verlag.

Weinberger, A., Ertl, B., Fischer, F., and Mandl, H. (2005). Epistemic and social scripts in computer-supported collaborative learning. *Instructional Science*, 33(1), 1-30.

Weinberger, A., Stegmann, K., Fischer, F., and Mandl, H. (2007). Scripting argumentative knowledge construction in computer-supported learning environments. In F. Fischer, I. Kollar, H. Mandl, and J. Haake (Eds.), *Scripting computer-supported communication of knowledge—Cognitive, computational and educational perspectives* (pp. 191-211). New York: Springer.

Weiss, I.R., Pasley, J.D., Smith, P.S., Banilower, E.R., and Heck, D.J. (2003). *Looking inside the classroom: A study of K-12 mathematics and science education in the United States.* Chapel Hill, NC: Horizon Research.

Wenger, E. (1998). *Communities of practice: Learning, meaning, and identity.* Cambridge, UK: Cambridge University Press.

West Virginia Department of Education. (2006). *21st century science K-8 content standards and objectives for West Virginia schools.* Available: http://wvde.state.wv.us/policies/csos.html [retrieved April 2009].

Wideen, M.F., Mayer-Smith, J., and Moon, B. (1998). A critical analysis of the research on learning-to-teach. *Review of Education Research*, 68(2), 130-178.

Wilson, C., Taylor, J., Kowalski, S., and Carlson, J. (2009). The relative effects of inquiry-based and commonplace science teaching on students' knowledge, reasoning and argumentation. *Journal of Research in Science Teaching* (accepted for publication July 2009). Available: http://www3.interscience.wiley.com/cgi-bin/fulltext/122686765/PDFSTART [retrieved January 2010].

Windschitl, M. (2009). *Cultivating 21st century skills in science learners: How systems of teacher preparation and professional development will have to evolve.* Paper prepared for the Workshop on Exploring the Intersection of Science Education and the Development of 21st Century Skills, National Research Council. Available: http://www7.nationalacademies.org/bose/WindschitlPresentation.pdf [retrieved June 2009].

Windschitl, M., and Thompson, J. (2006) Transcending simple forms of school science investigations: Can pre-service instruction foster teachers' understandings of model based inquiry? *American Educational Research Journal*, 43(4), 783-835.

REFERENCES

Windschitl, M., Thompson, J., and Braaten, M. (2009). *Fostering ambitious pedagogy in novice teachers: The new role of tool-supported analyses of student work*. Paper presented at the annual conference of the National Association of Research in Science Teaching, April, San Diego.

Wisconsin Department of Public Instruction. (2008). *Science applications: Performance standards G, Grade 8*. Madison: Author. Available: http://dpi.wi.gov/standards/scig8.html [retrieved April 2009].

Yoshida, M. (1999). *Lesson study: An ethnographic investigation of school-based teacher development in Japan*. Unpublished doctoral dissertation, University of Chicago.

Zimmerman, B.J. (2000). Attaining self-regulation: A social cognitive perspective. In M. Boekaerts, P.R. Pintrich, and M. Zeidner (Eds.), *Handbook of self-regulation* (pp. 451-502). San Diego: Academic Press.

Zimmerman, B.J. (2001). Theories of self-regulated learning and academic achievement: An overview and analysis. In B.J. Zimmerman and D.H. Schunk (Eds.), *Self-regulated learning and academic achievement: Theoretical perspectives*, 2nd ed. (pp. 1-38). Mahwah, NJ: Lawrence Erlbaum Associates.

Appendix A

Workshop Agenda and Participants

AGENDA

Exploring the Intersection of Science Education and the
Development of 21st Century Skills: A Workshop

February 5-6, 2009

Workshop Goals: The workshop is designed to explore six guiding questions, listed at the end of the agenda. Each workshop session focuses on one or more of these questions, as shown below.

Day 1: Thursday, February 5

8:00	Introductions (Working Breakfast)
8:30	**Welcoming Remarks**
	Carlo Parravano, Merck Institute for Science Education *Bruce Fuchs*, NIH Office of Science Education *Arthur Eisenkraft, Committee Chair*, University of Massachusetts, Boston
9:00	Introduction to KWL Activity: *Arthur Eisenkraft*, Moderator Carbonless copy notebooks will be used to record notes on what you know, want to know, and learned (KWL) throughout the workshop. You will be invited to share anonymous copies at the end of both days of the workshop.

The committee will use the copies to guide day 2 of the workshop, and staff may use the copies in the workshop summary report.

9:20   **Session 1: Panel Discussion on Demand for 21st Century Skills (Question 5)**

William Bonvillian, MIT, Washington, DC Office, Moderator

Emily DeRocco, The Manufacturing Institute, Panelist
Janis Houston, Personnel Decisions Research Institutes, Panelist
Ken Kay, Partnership for 21st Century Skills, Panelist

Guiding Questions for Session 1:
- How may development of 21st century skills through science education help prepare young people for lifelong learning, work, and citizenship (e.g., making personal decisions about health, making political decisions about global warming, making workplace decisions)?
- What is known about transferability of these skills to real workplace applications? What might have to change in terms of learning experiences to achieve a reasonable level of skill transfer?

10:15   Break

10:30   **Session 2: 21st Century Skills and Science Education Goals (Question 1)**

Marcia Linn, University of California-Berkeley, Moderator
Christian Schunn, University of Pittsburgh, Presenter
Bruce Fuchs, NIH Office of Science Education, Respondent

Guiding Questions for Session 2:
- What are the areas of overlap between 21st century skills and the skills and knowledge that are the goals of current efforts to reform science education?
- To what extent do science education standards treat science process skills and conceptual knowledge as separate or intertwined?
- What changes might be needed in science standards to support students' development of 21st century skills in the context of science education?

11:30   **Session 3: Adolescent Development of 21st Century Skills (Questions 2, 4)**
Christine Massey, University of Pennsylvania, Moderator

Eric Anderman, Ohio State University, Presenter
Gale Sinatra, University of Nevada, Las Vegas, Presenter
Susan Koba, Science Education Consultant, Respondent

Guiding Questions for Session 3:
- What is the state of research on children's and adolescents' developing ability to tackle complex tasks in the context of science education?
- What are the promising models or approaches for teaching these skills in science education settings? What, if any, evidence is available about the effectiveness of those models?

12:30   **Session 4: KWL Groups—Discussion of Sessions 1, 2, and 3 (Working Lunch)**
Room Assignments: 106-green; 202-red, 205-yellow; 104-blue; 202-light blue; 205-orange; 100-dark blue; 100-neon green. Participants, presenters, committee, and staff will break into small groups (assigned by color) to discuss what they learned during the morning sessions and what they want to know more about. Please plan to bring your notebook to your breakout session. Results will be shared in the plenary report out.

2:30   **Report Out** (Room 100)
William Sandoval, Moderator

3:30   **Break**

3:45   **Session 5: Promising New Science Curricula I (Questions 3 and 4)**
Arthur Eisenkraft, Moderator

Doug Clark, Arizona State University, Presenter
Rodger Bybee, Biological Sciences Curriculum Study (Emeritus), Presenter

Guiding Questions for Session 5:
- What unique, domain-specific aspects and practices of science appear to hold promise for developing 21st century skills?
- What are the promising models or approaches for teaching these skills in science education settings? What, if any, evidence is available about the effectiveness of those models?

5:00-5:15 Wrap-Up of the Day
Arthur Eisenkraft, Moderator

5:15-7:15 Reception for All Participants

Day 2: Friday, February 6

8:00 Review of the Previous Day's Activities (Working Breakfast)

8:30 Reflections on Science Education and 21st Century Skills
Carlo Parravano, Merck Institute for Science Education, Presenter

8:35 Session 6: Promising New Science Curricula II (Questions 3 and 4)
Carlo Parravano, Moderator

Janet Kolodner, Georgia Institute of Technology, Presenter
Joe Krajcik, University of Michigan, Presenter

Guiding Questions for Session 6:
- What unique, domain-specific aspects and practices of science appear to hold promise for developing 21st century skills?
- What are the promising models or approaches for teaching these skills in science education settings? What, if any, evidence is available about the effectiveness of those models?

10:00 Break

10:15 Session 7: Science Teacher Readiness for 21st Century Skills (Question 6)
William Sandoval, Moderator

Mark Windschitl, University of Washington, Presenter
Elizabeth Carvellas, NRC Teacher Advisory Council, Respondent

Guiding Questions for Session 7:
- What is known about how prepared science teachers are to help students develop 21st century skills?
- What new models of teacher education may support effective teaching and student learning of 21st century skills, and what evidence (if any) is available about the effectiveness of these models?

11:30 Break to Pick Up Lunch

APPENDIX A

11:45	**Session 8: KWL Groups—Discussion of Sessions 5, 6, and 7 (Working Lunch)** Room Assignments: 106-green; 109-red; 205-yellow; 205-blue; 109-light blue; 110-orange; 100-dark blue; 100-neon green. Participants, presenters, committee, and staff will break into small groups (assigned by color) of 9-12 per group to discuss what they learned in sessions 5, 6, and 7, and what they want to know more about. Please plan to bring your notebook to your breakout group. Results will be shared in the plenary report out.
12:50	*Report Out* (Room 100) *Christine Massey*, Moderator
1:15	**Session 9: Assessing 21st Century Skills (Questions 3 and 4)** *Marcia Linn*, Moderator *Janis Houston*, Personnel Decisions Research Institutes, Presenter *Maria Ruiz-Primo*, University of Colorado, Denver, Presenter *Guiding Questions for Session:* • What are the promising models or approaches for teaching these skills in science education settings? • What existing science assessments and other assessments hold promise for measuring 21st century skills and what evidence is available about these assessments? • What does a review of existing assessments suggest for design of future assessments to measure 21st century skills?
2:30	Break
2:45	**Session 10: Reflections on the Workshop by Committee Members** *Arthur Eisenkraft, Committee Chair*, University of Massachusetts, Boston *William Bonvillian*, MIT, Washington, DC Office *Marcia Linn*, University of California, Berkeley *Christine Massey*, University of Pennsylvania *Carlo Parravano*, Merck Institute for Science Education *William Sandoval*, University of California, Los Angeles
3:20	Closing Questions and Comments *Arthur Eisenkraft*, Moderator
3:30	Adjourn

Participants

Diane Adger-Johnson, National Institute of Allergies and Infectious Diseases
Susan Albertine, American Association of Colleges and Universities
Bernice Alston, National Aeronautics and Space Administration
Julie Angle, National Science Foundation
Raymond Bartlett, Teaching Institute for Excellence in STEM
Kirk Beckendorf, National Oceanic and Atmospheric Administration, Office of Education
Robert Bell, National Science Foundation
Mark Bloom, Biological Sciences Curriculum Study
Takiema Bunche Smith, Sesame Workshop
Michele Cahill, Carnegie Corporation of New York
Claudia Campbell, National Science Resources Center
Brian Carter, American Association for the Advancement of Science Fellow
Ines Cifuentes, American Geophysical Union
Charles Cox, U.S. Department of Labor
Hanna Doerr, National Commission on Teaching and America's Future
Janice Earle, National Science Foundation
Francis Eberle, National Science Teachers Association
Elizabeth K. Eder, Smithsonian American Art Museum
Curtis Ellis, House Committee on Education and Labor
Charles Fadel, Cisco Systems, Inc.
James Fey, National Science Foundation
Paul Ford, National Institutes of Health Office of Science Education
Jacob Foster, Massachusetts Department Education
John Hall, Pennsylvania Alliance for STEM Education
Peirce Hammond, U.S. Department of Education
Scott Jackson, National Institutes of Health
Sylvia James, National Science Foundation
Brian Jones, JBS International, Inc.
Jill Karsten, National Science Foundation
Michael Kaspar, District of Columbia Public Schools
John Kenny, Catholic University
Mary Kirchhoff, American Chemical Society
Miriam Lund, U.S. Department of Education
David Mandel, Carnegie Foundation-Institute for Advanced Study Commission on Mathematics and Science Education
Jacqueline Miller, Education Development Center
Zipporah Miller, National Science Teachers Association
Giuseppe (Pino) Monaco, Smithsonian Institution

Frank Niepold, National Oceanic and Atmospheric Administration, Climate Program Office
Shilpi Niyogi, Pearson Education
Douglas Oliver, National Science Foundation
Geraldine Otremba, Library of Congress
Eugene Owen, U.S. Department of Education
Stephen Provasnick, U.S. Department of Education
Linda Rosen, Education and Management Innovations, Inc.
Jim Rosso, Project Tomorrow
Gerhard Salinger, National Science Foundation
Gina Schatteman, National Institutes of Health
Reid Schwebach, National Research Council Board on Science Education
Jean Slattery, National Science Teachers Association
P. Gregory Smith, U.S. Department of Agriculture
Larry Snowhite, McGraw-Hill Education
Jaleh Soroui, American Institutes for Research
James Sylvan, National Science Foundation
Terri Taylor, American Chemical Society
Audrey Trotman, U.S. Department of Agriculture
Thomas Van Essen, Educational Testing Service
Susan Van Gundy, National Science Digital Library
Dave Vannier, National Institutes of Health Office of Science Education
Bill Watson, Smithsonian's National Museum of Natural History
Richard Weibl, American Association for the Advancement of Science
Brad Wiggins, U.S. Department of Labor, Employment and Training Administration
Joyce Winterton, National Aeronautics and Space Administration, Office of Education
Sarah Yue, National Science Foundation
Lee Zia, National Science Foundation

Appendix B

Biographical Sketches of Steering Committee Members, Presenters, Panelists, and Staff

STEERING COMMITTEE MEMBERS

Arthur Eisenkraft (*Chair*) is distinguished professor of science education and director of the Center of Science and Math in Context (COSMIC) at the University of Massachusetts, Boston. He is the lead author and the project director of Active Chemistry and Active Physics. His current research is associated with developing new models of professional development using distance learning, assessing technological literacy, and how to bring quality science instruction to all students, including those from traditionally underrepresented minorities. A fellow of the American Association for the Advancement of Science, he has received the Presidential Award for Excellence in Science Teaching (1986), the American Association of Physics Teachers' (AAPT) Distinguished Service Citation for "excellent contributions to the teaching of physics" (1989), Science Teacher of the Year, and the Disney American Teacher Award (1991). In 1999 he was elected president of National Science Teachers Association and was the sole recipient of an award for Excellence in Pre-College Physics teaching from AAPT. At the National Research Council, he has served on numerous panels, resulting in such diverse publications as the *National Science Education Standards*, *How People Learn: Bridging Research and Practice*, *Tech Tally*, *America's Lab Report: Investigations in High School Science*, and *Attracting Science and Mathematics Ph.D.s to Secondary School Education*. He has B.S. and M.S. degrees from Stony Brook University and a Ph.D. from New York University.

William Bonvillian is director of the Washington, DC, office of the Massachusetts Institute of Technology. Prior to this position, he served for 17 years as legislative director and chief counsel to U.S. Senator Joseph Lieberman. He is also an adjunct assistant professor at Georgetown University. He has written legislation in the areas of science, technology, and economic innovation and has an abiding interest in science and science education. Prior to leaving Senator Lieberman's office, he worked on legislation that came in direct response to the National Academies report *Rising Above the Gathering Storm: Energizing and Employing America for a Brighter Economic Future*. At the National Academies, he has been invited to speak to many groups about the legislative and policy process at the federal level and is a member of the Board on Science Education. He has a B.A. in history from Columbia University, an M.A.R. in religion from Yale University, and a J.D. from Columbia University School of Law.

Margaret Hilton (*Study Director*) is a senior program officer of the Board on Science Education. The workshop on science education and 21st century skills built on the workshop on future skill demands, which she directed in 2007. She is currently directing a review of the Occupational Information Network (O*NET). Previously, she has directed a study of high school science laboratories and contributed to workshops and studies of promising practices in undergraduate STEM; the role of state standards in K-12 education; foreign language and international studies in higher education; international labor standards; and the Information Technology workforce. Prior to coming to the NRC, at the Congressional Office of Technology Assessment, she directed studies of workforce training, work reorganization, and international competitiveness. She has a B.A. in geography, with High Honors, from the University of Michigan, a Master of Regional Planning degree from the University of North Carolina at Chapel Hill, and a Master of Human Resource Development degree from the George Washington University.

Marcia C. Linn is professor of development and cognition, specializing in education in mathematics, science, and technology, in the Graduate School of Education at the University of California, Berkeley. She also directs the Technology-Enhanced Learning in Science (TELS) center. She is a member of the National Academy of Education and a fellow of the American Association for the Advancement of Science, the American Psychological Association, and the Association for Psychological Science. Her board service includes the American Association for the Advancement of Science board, the Graduate Record Examination Board of the Educational Testing Service, the McDonnell Foundation Cognitive Studies in Education Practice board, and the Education and Human Resources Directorate at the National Science Foundation. She has written several books, including *Computers, Teachers, Peers* (2000), *Internet Environments for Science Education* (2004), and *Designing Coherent Science Education* (2008). She has received awards from the National Association for Research in Science Teaching and the Council of Scientific Society Presidents. At the National Academies, she has served on numerous committees: Support for Thinking Spatially: The Incorporation of Geographic Information Science Across the K-12 Curriculum; IT Fluency and High School Graduation Outcomes; the Board on Behavioral, Cognitive, and Sensory Sciences; and the Project on Information Technology Literacy. She has a B.A. in psychology and statistics and M.A. and Ph.D. degrees in educational psychology from Stanford University.

Christine Massey is director of research and education at the Institute for Research in Cognitive Science at the University of Pennsylvania. She is also the director of PENNlincs, an outreach arm of the institute, linking recent theory and research in cognitive science to education efforts in public schools and cultural institutions. She has directed a number of major collaborative research and development projects that combine research investigating students' learning and conceptual development in science and math with the development and evaluation of new curriculum materials, learning technology, and educational programs for students and teachers. She is also a primary participant in the Metromath Center for Math in America's Cities, a Center for Learning and Teaching, and the 21st Century Center for Cognition and Science Instruction. She was a Durant scholar and has a B.A. from Wellesley College and a Ph.D. in psychology with a specialization in cognitive development from the University of Pennsylvania. She is an Eisenhower fellow and has also been a fellow in the Spencer Foundation/National Academy of Education's postdoctoral fellowship program.

Carlo Parravano is executive director of the Merck Institute for Science Education. He is responsible for the planning, development, and implementation of numerous initiatives to improve science education. Previously he was professor of chemistry and chair of the Division of Natural Sciences at the State University of New York (SUNY) at Purchase. He is a fellow of the American Academy for the Advancement of Science and a national associate of the National Academies. He is a recipient of the SUNY Chancellor's Award for Excellence in Teaching, the National Science Teachers Association's Distinguished Service to Science Education Award, the Keystone Center's Leadership in Education Award, and Rutgers University's Distinguished Leader Award. He is a member of the National Academies' Board on Science Education (Executive Committee) and is principal investigator for a Mathematics/Science Partnership award from the National Science

Foundation. He has a B.A. in chemistry from Oberlin College and a Ph.D. in physical chemistry from the University of California, Santa Cruz.

William Sandoval is associate professor and head of the Division of Psychological Studies in Education in the Graduate School of Education and Information Studies at the University of California, Los Angeles. His teaching interests include the development of scientific reasoning, epistemologies of science and their effects on learning and teaching, technological supports for science inquiry, and technology as a transformative tool for instructional practice. His research interests focus on the development of scientific reasoning and inquiry skills, the design of technology-supported learning environments to support inquiry, and understanding and supporting effective inquiry teaching strategies. He was a key member of the BGuILE project and currently directs the Center for Embedded Networked Sensing Education Infrastructure project. He is a member of the editorial boards of the *Journal of the Learning Sciences*, *Science Education*, and *Cognition & Instruction*. At the National Academies, he was a member of the Committee on High School Science Laboratories. He has a B.S. in computer science from the University of New Mexico and a Ph.D. in learning sciences from Northwestern University.

PRESENTERS AND PANELISTS

Eric Anderman is professor of educational psychology at the Ohio State University. His research examines adolescent motivation. He has specifically studied and published about the transition from elementary school to middle school, the relation of motivation to academic cheating, and instructional interventions for HIV/pregnancy prevention for adolescents. He is associate editor of the *Journal of Educational Psychology* and a fellow of the American Psychological Association. He has a B.S. in psychology and Spanish from Tufts University, an Ed.M. in education from Harvard University, and M.A. and Ph.D. degrees in educational psychology from the University of Michigan.

Rodger W. Bybee is director emeritus of Biological Sciences Curriculum Study (BSCS). Previously, he was executive director of the National Research Council's Center for Science, Mathematics, and Engineering Education. At BSCS, he was principal investigator for four programs: an elementary school program entitled Science for Life and Living: Integrating Science, Technology, and Health; a middle school program entitled Middle School Science and Technology; a high school biology program titled Biological Science: A Human Approach; and a college program titled Biological Perspectives. He has been active in education for more than 30 years, having taught science at the elementary, junior and senior high school, and college levels. He has B.A. and M.A. degrees from the University of Northern Colorado and a Ph.D. in science education and psychology from New York University.

Betty Carvellas is the teacher leader for the Teacher Advisory Council at the National Academies. After teaching science for 39 years at the middle and high school levels, she retired in 2007. Her interests include interdisciplinary teaching, connecting school science to the real world, traveling with students on interdisciplinary field studies, and bringing inquiry into the science classroom. Her professional service includes work at the local, state, and national levels. She served as cochair of the education committee and was a member of the executive board of the Council of Scientific Society Presidents and is a past president of the National Association of Biology Teachers. In 2008, she was designated a national associate of the National Academies. She has a B.A. from Colby College, an M.S. from the State University of New York at Oswego, and a Certificate of Advanced Study from the University of Vermont.

Douglas Clark is assistant professor of science education and educational technology at Arizona State University in Tempe. His research focuses on issues of argumentation and conceptual change, often in the context of computer-enhanced learning environments. He is currently the principal investigator of an exploratory grant investigating physics learning in video games in terms of underlying game design, the design of representations and interfaces within games, and the structuring of social interactions outside games to optimize learning through discourse and argumentation. He recently completed a postdoctoral fellowship from the National Academy of Education and the Spencer Foundation analyzing students' knowledge structure coherence in physics in the United States, Mexico, China, Turkey, and the Philippines.

Emily S. DeRocco is president of the Manufacturing Institute and senior vice president of the National Association of Manufacturers (NAM). She oversees the education and research arm of the NAM and the design and operations of the new National Center for the American Workforce, which is dedicated to fostering a new generation of manufacturing workers for the 21st century. As assistant secretary of labor during the Bush administration, she was responsible for managing a $10 billion investment in the nation's workforce. She is a graduate of the Pennsylvania State University and has a J.D. from the Georgetown Law Center.

Bruce A. Fuchs is director of the Office of Science Education (OSE) at the National Institutes of Health (NIH). An immunologist who did research on the interaction between the brain and the immune system, Fuchs was previously a researcher and a teacher on the faculty of the Medical College of Virginia. Currently he is responsible for monitoring a range of science education policy issues and providing advice to NIH leadership. He also directs the creation of the NIH Curriculum Supplement Series, which highlights the medical research findings of NIH and is designed to meet teacher's educational goals as outlined in the *National Science Education Standards*. He has a B.S. in biology from the University of Illinois and a Ph.D. in immunology from Indiana State University.

Janis S. Houston is vice president and principal research scientist at Personnel Decisions Research Institutes in Minneapolis. She has directed, codirected, or been the lead research team member on numerous consulting and research projects, in both the public and private sectors, most of which have involved personnel selection or employee development and performance measurement. She has helped clients design their employee development programs and structure their career tracking and planning efforts. She has led projects to develop and validate selection and promotion tools and designed and implemented a number of test administration systems involving test sites across the country and overseas. She has developed a number of programs, such as for training and coaching individuals to conduct selection and promotion interviews. She has an M.A. in industrial psychology from the University of Minnesota.

Kenneth Kay is chief executive officer and cofounder of the e-Luminate Group in Tucson, Arizona. He also serves as president of the Partnership for 21st Century Skills. He has been primarily concerned with defining the potential and promoting the importance of information technology applications in education, health care, electronic commerce, and government services. As executive director of the CEO Forum from 1996 to 2001, he facilitated dialogue between leaders in the business, government, and education fields and led the group through development of the StaR Chart (School Technology & Readiness Guide), used by schools across the country to make better use of technology in the classroom. In 1989-2003, Kay was the founding executive director of the Computer Systems Policy Project, a CEO advocacy group for information technology policy. He has a law degree from the University of Denver and an undergraduate degree from Oberlin College.

Susan Koba is a science education consultant working primarily with the National Science Teachers Association on their e-Learning Center. She recently retired after more than 30 years in the Omaha Public Schools, teaching for over 20 years in middle and high school and then serving as a district mentor and leader. She led the development of an online teacher professional development environment and coordinated professional development in science and mathematics for 60 schools during her service as project director and professional development coordinator for the district's Urban Systemic Program. She has published and presented on various topics, including school and teacher change, equity in science, inquiry, and action research. She has a B.S. in biology and secondary education from Doane College, an M.A. in biology from the University of Nebraska–Omaha, and a Ph.D. in science education from the University of Nebraska–Lincoln.

Janet Kolodner is regents' professor in the College of Computing at the Georgia Institute of Technology. She pioneered the computer reasoning method called case-based reasoning, a way of solving problems based on analogies to past experiences, and her lab emphasized case-based reasoning for situations of real-world complexity. For the past decade, she has focused on using the model to design science curriculum for middle school and more recently has applied it to informal education—after-school programs, museum programs, and museum exhibits. She was the founding director of Georgia Tech's EduTech Institute, whose mission is to use what is known about cognition to inform the design of educational technology and learning environments. She also served as coordinator of Georgia Tech's cognitive science program for many years. She has a B.S. in mathematics and computer science from Brandeis University and M.S. and Ph.D. degrees in computer science from Yale University.

Joseph S. Krajcik is professor of science education and associate dean for research in the School of Education at the University of Michigan–Ann Arbor. Working with teachers in science classrooms to bring about sustained change, he aims to create classrooms that focus on students collaborating to find solutions to important intellectual questions based on essential learning goals and use new learning technologies as productivity tools. He codirects the Center for Highly Interactive Classrooms, Curriculum and Computing in Education at the University of Michigan and is a coprincipal investigator in the Center for Curriculum Materials in Science and the National Center for Learning and Teaching Nanoscale Science and Engineering. He taught high school chemistry for seven years in Milwaukee, Wisconsin. He has a Ph.D. in science education from the University of Iowa.

Maria Araceli Ruiz-Primo is associate professor in the School of Education and Human Development at the University of Colorado, Denver, and also director of the Research Center and director of the Laboratory for Educational Assessment, Research, and Innovation (LEARN). She specializes in educational assessment. Her research work focuses on the development and technical evaluation of innovative science learning assessment tools—including performance tasks, concept maps, and student science products—and on the development of a conceptual framework of academic achievement. She has conducted research on the instructional sensitivity of assessments and their proximity to the enacted curriculum. She participated in the development of the science teacher certification assessment of the National Board for Professional Teaching Standards and in the development and evaluation of teacher enhancement programs for elementary science teachers. She is the first author of the *Student Guide, Statistical Reasoning for the Behavioral Sciences*.

Christian Schunn is associate professor of psychology with appointments in the Intelligent Systems Program and the Learning Sciences and Policy program at the University of Pittsburgh. He is also a research scientist at the Learning Research and Development Center. His research generally examines the cognitive foundation of intelligent behavior and tools that further support its acquisition and deployment. Specific research areas include engineering innovation, scientific reasoning, and science education. He codirected a large Math Science Partnership from 2003 to 2006, and now codirects the Center for e-Design and the Institute for Education Science's 21st Century Center for Cognition and Science Education. He has a Ph.D. in psychology from Carnegie Mellon University.

Elena Silva is senior policy analyst at Education Sector in Washington, DC, where she oversees the organization's teacher quality work and develops and directs mixed-method research projects designed to evaluate education reform efforts at the national, state, and local levels. Her recent publications include *Waiting to Be Won Over: Teachers Speak on the Profession, Unions, and Reform* (2008) and *The Benwood Plan: A Lesson in Comprehensive Teacher Reform* (2008). Silva serves as a member of the design team for the National Center for Education Statistics' Schools and Staffing Survey and the Teacher Follow-Up Survey. Prior to joining Education Sector, Silva was the director of research for the American Association of University Women. She has M.A. and Ph.D. degrees (the latter in education) from the University of California, Berkeley.

Gale M. Sinatra is professor of educational psychology at the University of Nevada at Las Vegas. Her model of conceptual change learning emphasizes the role of motivation in conceptual change. Her book (with Paul Pintrich) *Intentional Conceptual Change* examines the students' role in facilitating their own knowledge change. Her recent article, "The Warming Trend in Conceptual Change Research" (2005) describes the new "hot cognition" view of conceptual change inspired by Pintrich's work. She is currently co-principal investigator of a grant exploring the challenges of teaching and learning about biological evolution in the United States, which include emotional and motivational barriers. She has B.S., M.S., and Ph.D. degrees in psychology with a minor in educational measurement from the University of Massachusetts, Amherst.

Mark Windschitl is associate professor of science education at the University of Washington–Seattle. He is the head of the Teachers' Learning Trajectories Initiative, which is currently working on two fronts regarding the early career development of science teaching expertise. One track includes longitudinal studies of science educators who are able take up ambitious, equitable, and effective forms of teaching over time. These investigations are aimed at developing theory around how and why early career educators develop pedagogical expertise across specific learning-to-teach contexts (in university course work, during student teaching, in professional learning communities, and during the first two years in the classroom). The other track of inquiry is a five-year project to develop and study a system of tools and tool-based practices for early career and preservice secondary science teachers that support transitions from novice to expert-like pedagogical reasoning and practice. He has a Ph.D. from Iowa State University.